D1794819

# 1 MONTH OF FREE READING

at

## www.ForgottenBooks.com

By purchasing this book you are eligible for one month membership to ForgottenBooks.com, giving you unlimited access to our entire collection of over 1,000,000 titles via our web site and mobile apps.

To claim your free month visit:

www.forgottenbooks.com/free1113524

ISBN 978-0-331-37271-7
PIBN 11113524

# NBS TECHNICAL NOTE **666**

# NATIONAL BUREAU OF STANDARDS

The National Bureau of Standards [1] was established by an act of Congress March 3, 1901. The Bureau's overall goal is to strengthen and advance the Nation's science and technology and facilitate their effective application for public benefit. To this end, the Bureau conducts research and provides: (1) a basis for the Nation's physical measurement system, (2) scientific and technological services for industry and government. (3) a technical basis for equity in trade, and (4) technical services to promote public safety The Bureau consists of the Institute for Basic Standards, the Institute for Materials Research, the Institute for Applied Technology, the Institute for Computer Sciences and Technology, and the Office for Information Programs.

THE INSTITUTE FOR BASIC STANDARDS provides the central basis within the United States of a complete and consistent system of physical measurement; coordinates that system with measurement systems of other nations; and furnishes essential services leading to accurate and uniform physical measurements throughout the Nation's scientific community, industry, and commerce. The Institute consists of a Center for Radiation Research, an Office of Measurement Services and the following divisions:

Applied Mathematics — Electricity — Mechanics — Heat — Optical Physics — Nuclear Sciences [2] — Applied Radiation [2] — Quantum Electronics [3] — Electromagnetics [3] — Time and Frequency [3] — Laboratory Astrophysics [3] — Cryogenics [3].

THE INSTITUTE FOR MATERIALS RESEARCH conducts materials research leading to improved methods of measurement, standards, and data on the properties of well-characterized materials needed by industry, commerce, educational institutions, and Government; provides advisory and research services to other Government agencies; and develops, produces, and distributes standard reference materials The Institute consists of the Office of Standard Reference Materials and the following divisions:

Analytical Chemistry — Polymers — Metallurgy — Inorganic Materials — Reactor Radiation — Physical Chemistry

THE INSTITUTE FOR APPLIED TECHNOLOGY provides technical services to promote the use of available technology and to facilitate technological innovation in industry and Government. cooperates with public and private organizations leading to the development of technological standards (including mandatory safety standards), codes and methods of test; and provides technical advice and services to Government agencies upon request. The Institute consists of a Center for Building Technology and the following divisions and offices:

Engineering and Product Standards — Weights and Measures — Invention and Innovation — Product Evaluation Technology — Electronic Technology — Technical Analysis — Measurement Engineering — Structures, Materials, and Life Safety [4] — Building Environment [4] — Technical Evaluation and Application [4] — Fire Technology.

THE INSTITUTE FOR COMPUTER SCIENCES AND TECHNOLOGY conducts research and provides technical services designed to aid Government agencies in improving cost effectiveness in the conduct of their programs through the selection, acquisition, and effective utilization of automatic data processing equipment; and serves as the principal focus within the executive branch for the development of Federal standards for automatic data processing equipment, techniques, and computer languages. The Institute consists of the following divisions:

Computer Services — Systems and Software — Computer Systems Engineering — Information Technology.

THE OFFICE FOR INFORMATION PROGRAMS promotes optimum dissemination and accessibility of scientific information generated within NBS and other agencies of the Federal Government; promotes the development of the National Standard Reference Data System and a system of information analysis centers dealing with the broader aspects of the National Measurement System; provides appropriate services to ensure that the NBS staff has optimum accessibility to the scientific information of the world. The Office consists of the following organizational units:

Office of Standard Reference Data — Office of Information Activities — Office of Technical Publications — Library — Office of International Relations

---

[1] Headquarters and Laboratories at Gaithersburg, Maryland, unless otherwise noted, mailing address Washington, D C. 20234.
[2] Part of the Center for Radiation Research.
[3] Located at Boulder, Colorado 80302.
[4] Part of the Center for Building Technology.

# Efflux of Gaseous Hydrogen or Methane Fuels from the Interior of an Automobile

J. M. Arvidson

J. Hord

D. B. Mann

Cryogenics Division
Institute for Basic Standards
National Bureau of Standards
Boulder, Colorado   80302

Technical note no. 666

U.S. DEPARTMENT OF COMMERCE, Frederick B. Dent, Secretary

NATIONAL BUREAU OF STANDARDS Richard W Roberts Director

Issued March 1975

Library of Congress Catalog No. 75-600005
National Bureau of Standards Technical Note 666

Nat Bur. Stand (U.S.), Tech Note 666, 56 pages (Mar. 1975)
CODEN: NBTNAE

For sale by the Superintendent of Documents, U S Government Printing Office, Washington, D C 20402
(Order by SD Catalog No C13 46 666) $1 10

CONTENTS

## LIST OF FIGURES

iii

iv

LIST OF TABLES

EFFLUX OF HYDROGEN OR METHANE GASES FROM THE INTERIOR OF

A GASEOUS-FUELED AUTOMOBILE

J. M. Arvidson, J. Hord and D. B. Mann

ABSTRACT

Gasoline-powered automobiles are being converted to operate on gaseous fuels such as $H_2$ or $CH_4$. These fuels are commonly stored in containers located in the trunk of the car. Potential leakage of these gaseous fuels into the passenger compartment of the vehicle constitutes a safety threat. Definitive experiments were performed to identify the explosion hazards, establish venting criteria and obviate general safeguards for $H_2$ or $CH_4$ fueled passenger vehicles. Appropriately designed ventilation systems significantly reduce the safety hazards associated with accumulated combustible gases. Vents are recommended for all autos _converted_ to burn $H_2$ or $CH_4$ and may possibly be eliminated in new cars that are _designed_ for gaseous fuel operation. Combustible gas warning systems are recommended, at least in the interim, for all (converted and new-design) gaseous fueled vehicles. $H_2$ and $CH_4$ gases appear equally safe as vehicular fuels if used in properly designed vehicles.

Key words: Automobile; detection; dispersion; explosion; fire; hydrogen; leakage; methane; safety; vents.

1.0 SUMMARY

Conversion of gasoline-powered vehicles to operate on gaseous fuels is becoming increasingly popular. One safety hazard with such conversions is the potential for combustible gas to leak into the passenger compartment from its storage container in the trunk of the car. Controlled experiments were performed to define fire and explosion hazards that may result from this leakage. The rate of mixing within and efflux of combustible gases ($H_2$ or $CH_4$) from the passenger compartment of an American passenger vehicle were determined.

* This study was performed at the National Bureau of Standards and was sponsored, in part, by the General Services Administration.

These gas dispersion characteristics were found to depend heavily on gas in-leakage flow rates and ventilation of the passenger compartment. Tests were conducted at injection leakage flow rates of 19 to 180 sccs (standard cubic centimeters per second) and with roof (and firewall) vent areas of 0.0 to 129 $cm^2$. All tests were performed with the car doors and windows closed in both the vented and non-vented configurations. Combustible gas sensors were judiciously mounted within the car interior to provide the desired dispersion data.

The test data clearly show the safety advantages of well-ventilated passenger compartments. Roof (and firewall) vents appear highly desirable in autos converted to operate on gaseous fuels, even if a membrane is installed to isolate the fuel system and passenger compartments. Appropriate vents delay, and for low flows prevent, flammable concentrations of $H_2$ or $CH_4$ in the passenger compartment. These vents also accelerate dilution of flammable gas mixtures. Venting criteria and accompanying personnel risks are developed herein.

Hydrogen disperses more rapidly than methane but for the same volumetric leakage flow rate identical vent areas are required to avoid an incipient fire hazard in the car interior. Both gases appear equally safe if used in a properly designed vehicle. Gaseous dispersion appeared to occur mainly by buoyancy-induced flow.

It is believed that ventilation and combustible gas warning systems are highly desirable equipment for all passenger vehicles converted to burn $H_2$ or $CH_4$. Autos designed for $H_2$ or $CH_4$ fuels could conceivably omit the ventilation system by incorporating appropriate integral design features as commonly employed in gasoline-powered vehicles. The combustible gas detector system is recommended for all gaseous fueled cars until more operating experience is acquired.

## 2.0  INTRODUCTION

With increasing emphasis on air quality, clean-burning fuels such as hydrogen and methane make prime candidates for use in internal combustion engines [1-10]. In addition to reduced emissions a cost savings may be realized in using gaseous fuels [9,10]. This economic advantage is primarily due to the soaring price of gasoline and the absence of a tax schedule on natural gas as a motor fuel. Taxation, scarcity of gaseous fuels and concomitant increasing gas prices could eliminate the economic incentive [10] to convert gasoline-powered vehicles to gaseous fuel operation. Exhaust emissions are greatly reduced with $H_2$ or $CH_4$ fuels; $H_2$-fueled autos [11] already satisfy 1975-1976 Federal emission standards while $CH_4$-fueled vehicles [10,18] apparently do not quite meet these rigid specifications.

Due to increasing interest in gaseous fueled vehicles and numerous fleet conversions [8,10] the safety of gaseous fueled vehicles has come under closer scrutiny. The Department of Transportation [12,13], Massachusetts Turnpike Authority [10,14], state of California [15,16], Compressed Gas Association [15], and General Services Administration (this study and reference [9]) have initiated research and/or proposed standards and regulations for the safe operation of such vehicles. Definitive vehicle impact test data are available [12] and safety criteria for gaseous vs gasoline-fueled vehicles have been developed [10]. To perform a thorough analysis of fire and explosion hazards associated with alternate fuels it is necessary to acquire mixing and dispersion characteristics of candidate gaseous fuels. Such data are necessary to evaluate the hazards associated with vehicle smash-ups and with inadvertant accumulation of fuel gases inside of the vehicle. The latter is far more subtle and is believed to be the cause of a nonfatal incident involving a natural-gas fueled auto. Fuel apparently leaked into the car interior and was ignited while starting the engine.

It was this event that provided impetus for the study reported herein. The objective of this experiment is to determine the mixing and efflux characteristics of hydrogen and methane when injected (leaked) into the passenger compartment of a vehicle. The vehicle used for this investigation was an American manufactured dual-fuel (natural gas/gasoline), 1970 sedan. This car, furnished by the General Services Administration, is similar to the one that accidentally exploded as described above. The dual-fuel car was modified to operate on either gasoline or natural gas. The natural gas is stored as a liquid (at ~ 112 K) inside a 64 liter (17 gallon) dewar located in the trunk -- with given weight or volume constraints, hydrogen or methane in their liquid states provide a much greater range for the vehicle than if stored as a gas.

### 3.0 EXPERIMENTAL PLAN

In this section we discuss the rationale used to design our experiment. We wish to simulate the leakage of gaseous fuel from the trunk to the passenger compartment of the auto. Most gaseous-fuel conversion kits for gasoline-powered autos use the trunk space for storage, refueling, and venting operations. In some cases efforts are made to seal-off the trunk from the passenger compartment and both trunk and passenger spaces may be vented. Vent gas is normally ducted outside of the trunk to the rear of the vehicle. Gas can conceivably leak into the passenger compartment in a variety of ways: a defective, damaged, or nonexistent trunk sealing membrane could permit entry of gas from a defective, neglected or damaged fuel storage system, e.g., the result of a collision, loose connections, malfunctioning valves or regulators, improper design/construction/inspection, abuse, etc.

To cover a variety of these potential hazards we attempted to answer the following questions: (a) What are reasonable inflow (leakage) rates, cumulative leakage quantities and leakage injection locations? (b) What are experimentally reasonable gas residence times in the passenger compartment? (c) What is the effect of increased roof vent area? (d) What is the effect of leakage gas temperature and environmental temperature on test results? (e) What is the effect of wind direction and velocity on the rate of efflux of gaseous fuels from the car interior? and (f) What instrumentation is required?

(a) Inflow (leakage): It was arbitrarily decided that the gas ($H_2$ or $CH_4$) would be injected into the passenger compartment just ahead of the rear window at the rear parcel shelf. The choice of this entry location is

4

slightly defensible as both $H_2$ and $CH_4$ are lighter than air and tend to rise to the top of the trunk compartment where the parcel shelf is located. Leakage flow rates were also arbitrarily selected to range from 19 to 180 sccs (0.04 to 0.38 scfm). These leakage rates may be slightly high for most gaseous fuel installations but are considered reasonable. In practice, the leakage rates will be governed by the sealing quality of the trunk membrane. Higher or lower leakage rates within the trunk space are certainly possible but it is assumed that trunk vents would not permit membrane leakage rates in excess of 180 sccs. Accordingly, only a trunkful (0.216 $m^3$) of combustible gas was leaked into the passenger compartment during a single test. This quantity of gas, while arbitrarily chosen, is also conducive to experimenter safety because it limits a homogeneous mixture of $H_2$-air in the car interior to concentrations below the lower detonable limit (LDL).

(b) Gas residence times: The leakage flow rates and roof vent area, as discussed in the next few paragraphs, are integrally related to the residence time of the combustible gas within the passenger compartment. From both practical and experimental viewpoints it was felt that residence times should not exceed several hours. Preliminary experimental data indicated that this criterion could be met with the chosen leakage flows and roof vent capacities.

(c) Roof vent area: Injection of gaseous fuel causes expulsion of air, and subsequently fuel, from the passenger compartment of the car. Fuel escapes from the car through the roof vent and a multitude of smaller openings by buoyant and diffusion flow mechanisms. For identical leakage paths, flow modes, etc., $H_2$ or $CH_4$ will expel air with an exit velocity, $V_e \propto \sqrt{\Delta P_B}$; $\Delta P_B$ = buoyant pressure (density) gradient. Thus, $V_e \propto C_o \sqrt{\rho_A - \rho_f}$, where $C_o$ is approximately constant if a fixed volume of fuel gas is admitted to the car interior -- $C_o$ is numerically equivalent to the square root of the fuel volume-to-area ratio, $\rho_A$ = air density and $\rho_f$ = fuel density. Using the foregoing expression, buoyant forces will favor the rapid efflux of $H_2$ gas by the ratio

$$\frac{(V_e)_{H_2}}{(V_e)_{CH_4}} \propto \sqrt{\frac{\rho_A - \rho_{H_2}}{\rho_A - \rho_{CH_4}}} \approx 1.45.$$

Diffusion velocities also favor $H_2$ dispersion by the ratio of $\approx 0.63/0.20 \approx$ 3:1; however, diffusion flow rates are nearly negligible in comparison with density (buoyancy) displacement rates.

While $H_2$ should escape more readily than $CH_4$ from a specific auto interior it must also inflow (leakage through trunk membrane) more rapidly. The leakage of $H_2$ vs $CH_4$ into the car through a fixed geometry leak path favors $CH_4$; volumetric leakage flow [17] of $H_2$ gas will be 1.25 (viscous flow) to 3 times as large as $CH_4$ leakage (at identical temperatures and pressures). Thus, it appears that $H_2$ could leak into the car 1.25 to 3 times as fast as $CH_4$ and could leak out again 1.45 to 3 times as fast as could $CH_4$.

Private consultants to the DoT (no source or reference is traceable) have reportedly suggested that natural gas fueled cars be equipped with 1 in$^2$ of roof vent area per 36 ft$^3$ of car volume. Excluding trunk volume the experimental car volume $\approx 140$ ft$^3$; therefore, a vent area of $\approx 4$ in$^2$ would be prescribed. By performing buoyant flow calculations we verified that a 4 in$^2$ roof vent area is reasonable; a volume of air equivalent to a trunkful of air could be expelled through the vent in less than 5 minutes. These calculations must be considered rough estimates as the fuel concentration (and consequently fuel efflux) in the passenger compartment is constantly changing during a test. Also, these estimates are valid only if air and fuel expelled from the car are replaced with make-up air at the same flow rate, i.e., inleakage of air around doors, windows, floor boards, etc., must be sufficient to replace air and fuel expelled through the roof vent. Otherwise, the air inleakage capacity (dependent upon vapor-tightness of the car) will govern, or at least influence, the rate at which the air-fuel mixture can escape through the roof vent. This consideration prompted us to provide a fresh-air vent through the car firewall near the bottom of the passenger compartment. In all vented tests the firewall vent area was identical to the roof vent area. Tests were also conducted with the auto totally closed — no vents — to determine "worst case" results. Many autos already feature fresh-air ducts and they could be easily installed at low cost for use in favorable climates.

Our confidence in the efflux computations was rather low because of 1) the uncertainty of existing flow paths into and out of the car and 2) the lack of information on buoyant mixing and purging characteristics. Thus, it was decided that firewall and roof vent areas should be varied to determine their true effect on gas residence times and efflux rates.

(d) & (e) Environmental Conditions: In preliminary experiments with natural gas the temperature of the injected gas was varied from 112 to 300 K with nearly identical mixing and dispersion results. On the basis of these tests we decided to inject the gaseous fuels at room temperature and simply measure the temperature of the air-fuel mixture in the passenger compartment. Environmental temperature was also controlled near room temperature by placing the auto and experimental equipment inside of a closed building, designed for hazardous experiments. By locating the auto indoors the random influence of changing winds (on efflux rates) was also eliminated. Our preliminary tests indicated that changing winds were by far the most important environmental consideration -- particularly in those tests that required several hours for completion and especially in Boulder, Colorado.

(f) Essential Instrumentation: As previously explained, measurement of gas injection temperature was not necessary and environmental temperature was held nearly constant. The barometric pressure ($\sim$ 0.83 bar) does not vary enough to affect efflux rates or justify concern. Air-fuel temperature inside of the passenger compartment was continuously measured using four strategically located thermocouples. Pressure difference between the passenger compartment and auto environment was measured with a water-filled manometer. Gaseous fuel concentration was continually sensed at eight specified locations within the passenger compartment using combustible gas detectors. Gas injection flow rates and integrated (total) flow were also measured.

### 4.0 EXPERIMENTAL APPARATUS

The apparatus consisted of the experimental auto, appropriately instrumented and situated within a properly equipped laboratory building.

### 4.1 EXPERIMENTAL FACILITY

The test vehicle was located in the test bay of a metal building equipped with explosion-proof electrical fixtures and a roof-mounted explosion-proof exhaust fan. Large access doors to the test bay were sealed off with 0.015 cm thick plastic sheets to provide pressure relief for the building in the event of an explosion in the auto, see figure 1. The roof fan was operated and all doors and windows in the building were kept closed during a test; therefore, hydrogen (or methane) exiting from the test vehicle escaped from the building through the roof fan. A continuous and uniform flow of air into the building and out through the roof fan is assured by the 'leaky' construction of this type of prefabricated building.

Experimental control was performed in a partitioned room adjacent to the test bay. Two steel plates (1/4-inch x 4 ft x 8 ft) were strapped together to form a protective barrier 8 ft x 8 ft. This barrier was secured to the partition separating the control room and the test bay.

8

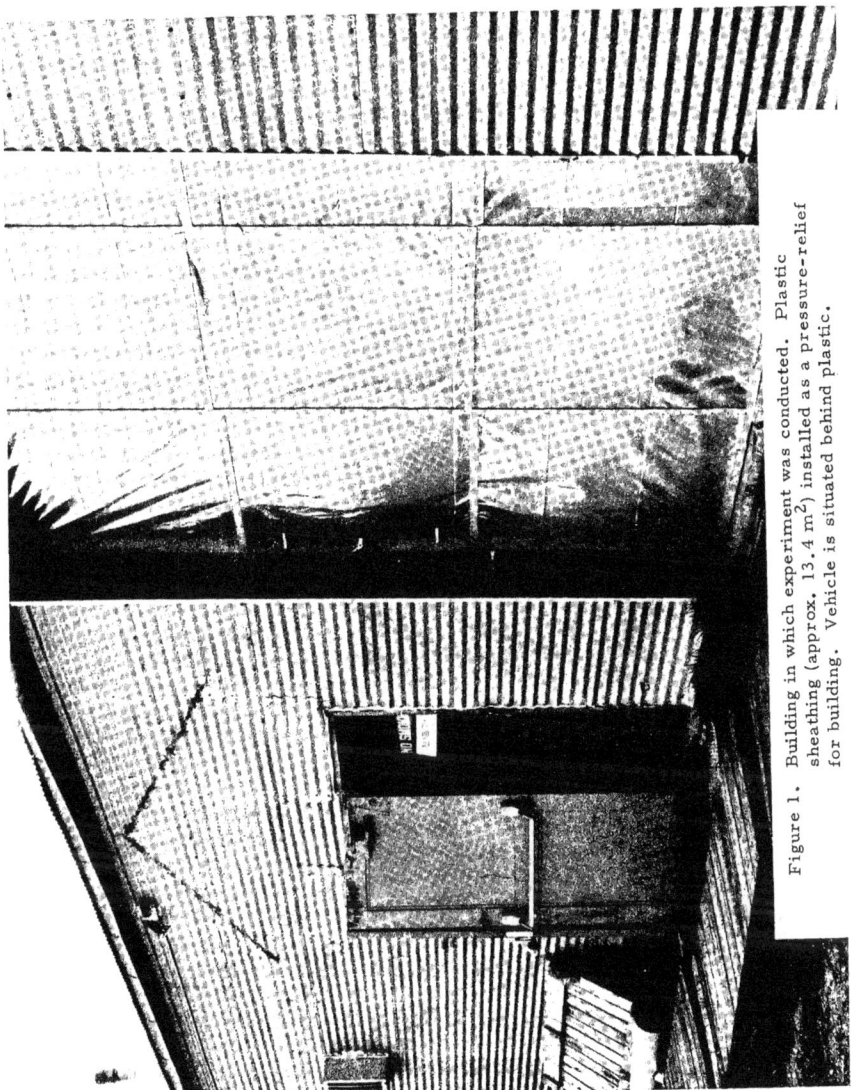

Figure 1. Building in which experiment was conducted. Plastic sheathing (approx. 13.4 $m^2$) installed as a pressure-relief for building. Vehicle is situated behind plastic.

9

## 4.2 EXPERIMENTAL VEHICLE

The test vehicle was a standard 1970 sedan manufactured in the United States. All doors and windows were in their normal closed positions during all tests. The engine cowling and trunk lids were both kept in the 'up' (open) positions during all tests. An unobstructed vent of 129 $cm^2$ (20 $in^2$) area was cut in the center of the roof and an identical opening was cut into the firewall. The firewall vent was ducted to the bottom center of the vehicle just ahead of the front seat, see figures 2 and 3. Smaller roof and firewall vent areas were obtained by covering a portion of the 129 $cm^2$ areas. The rear seat and parcel shelf were removed and aluminum panels (0.076 cm thick) were installed to isolate the trunk area from the passenger compartment. The aluminum sheets were held in place with sheet metal screws and the edges were sealed with heating-and-ventilating duct tape. The rear seat was then replaced.

The system used for introducing gas ($H_2$ or $CH_4$) into the car and for maintaining a precise flow rate is shown in figure 4. Flow modulation was accomplished by throttling with valve 4 or valve 5 and monitoring flowmeter 1 (FM1) or flowmeter 2 (FM2). The vehicle was equipped with three equally-spaced (0.317 cm dia.) copper tubes near the base of the rear window; these gas inlet tubes were of equal length and were manifolded to a supply line (0.635 cm dia.) connected to the control panel.

A nitrogen purge line was used to inert the passenger compartment upon conclusion of a test, see figures 2 and 5. The check valve located near the tee-outlet was installed to prevent the possible back-flow of combustible gas into the purge line.

Finally, the vehicle was instrumented as described in the following section.

## 4.3 INSTRUMENTATION DETAILS

During an entire test the pressure inside of the passenger compartment, relative to pressure on the exterior of the vehicle, was monitored

10

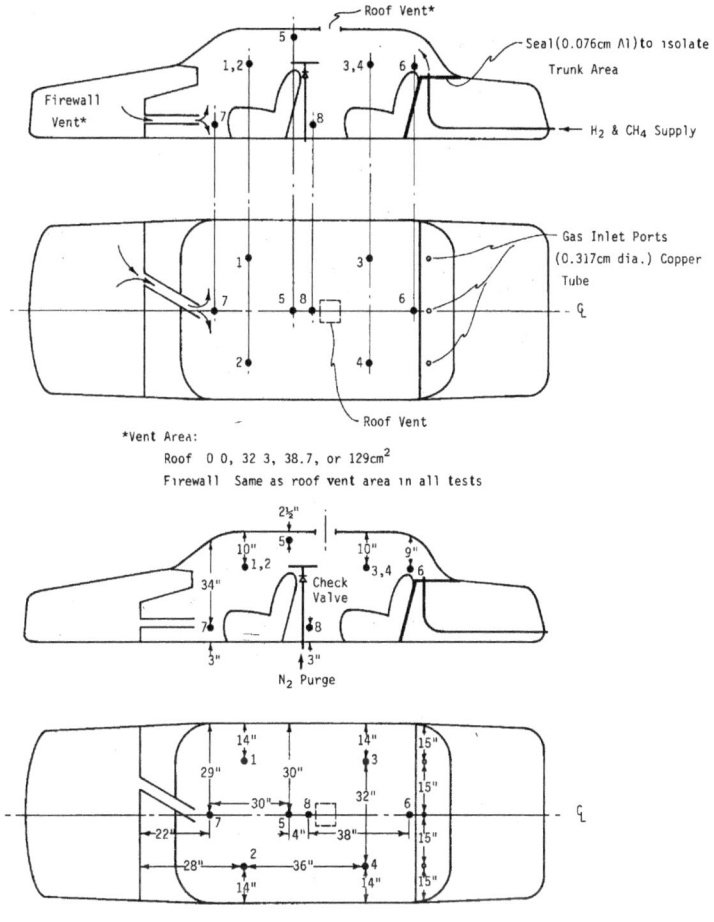

Figure 2. Schematic of sensor (1-8) and thermocouple (5-8) locations.

11

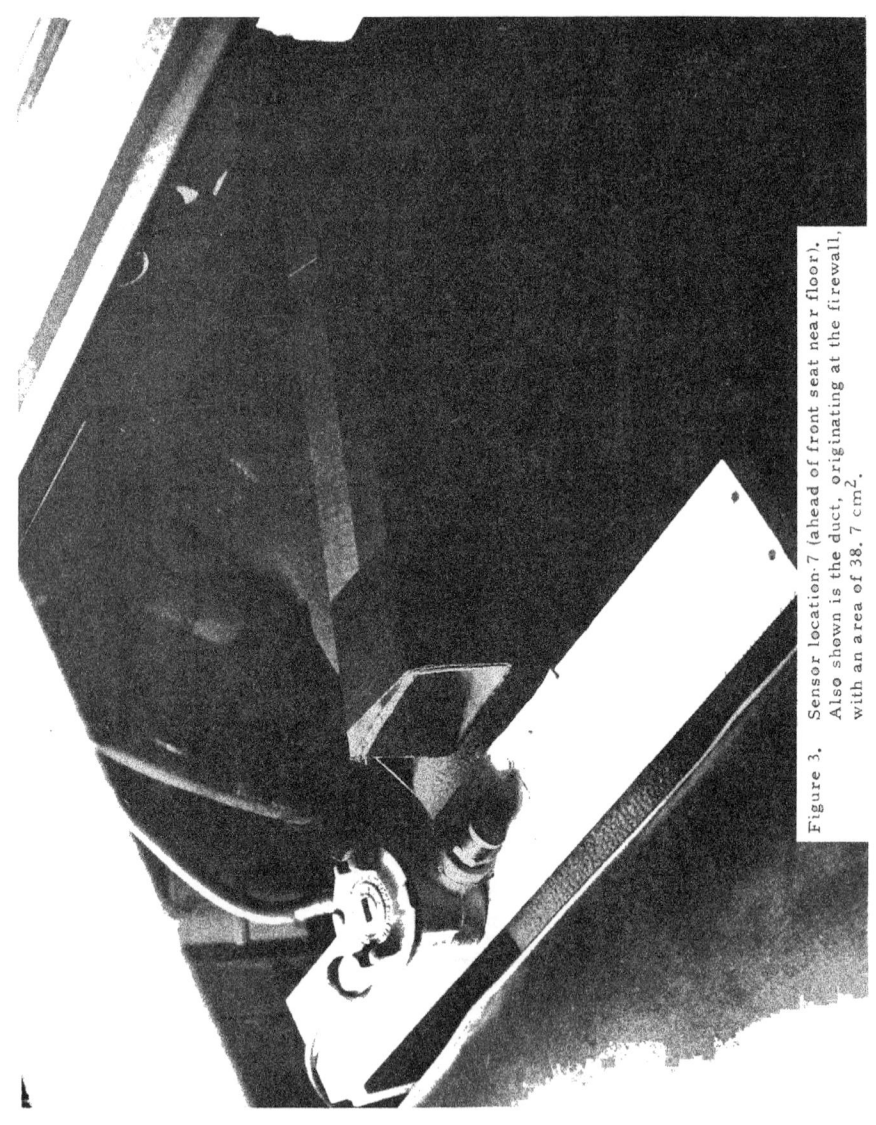

Figure 3. Sensor location 7 (ahead of front seat near floor). Also shown is the duct, originating at the firewall, with an area of 38.7 cm$^2$.

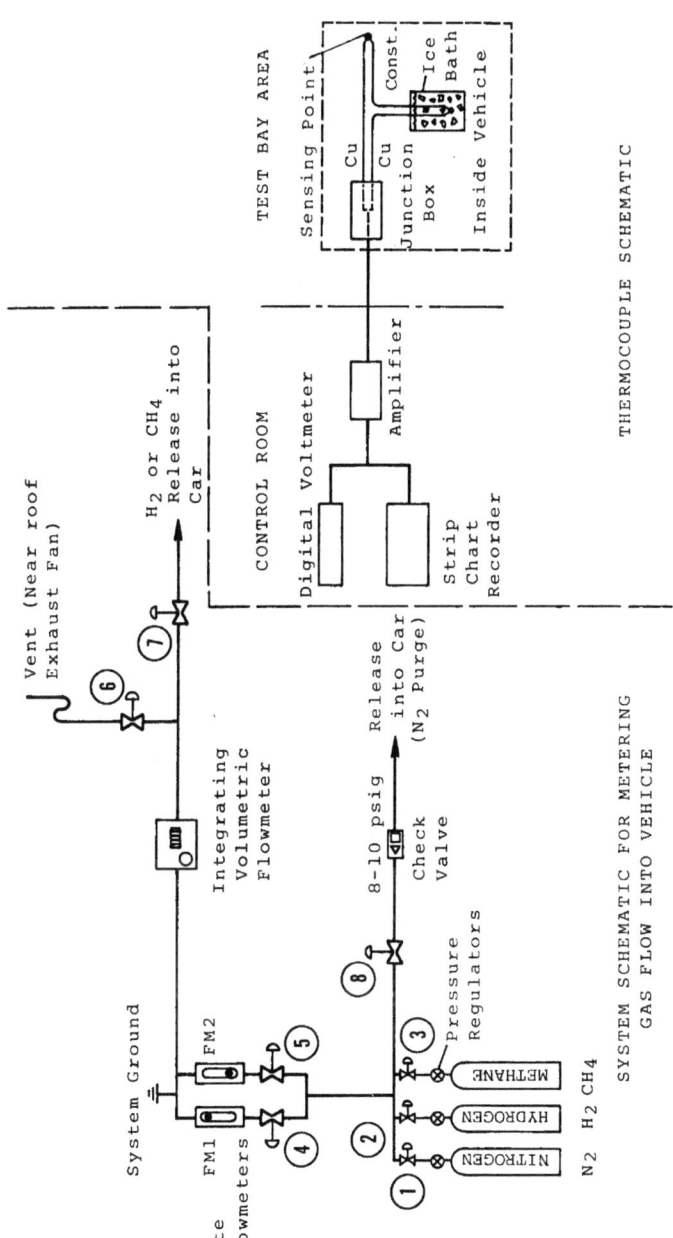

Figure 4. System schematic for metering gas flow into vehicle and typical thermocouple measuring circuit.

13

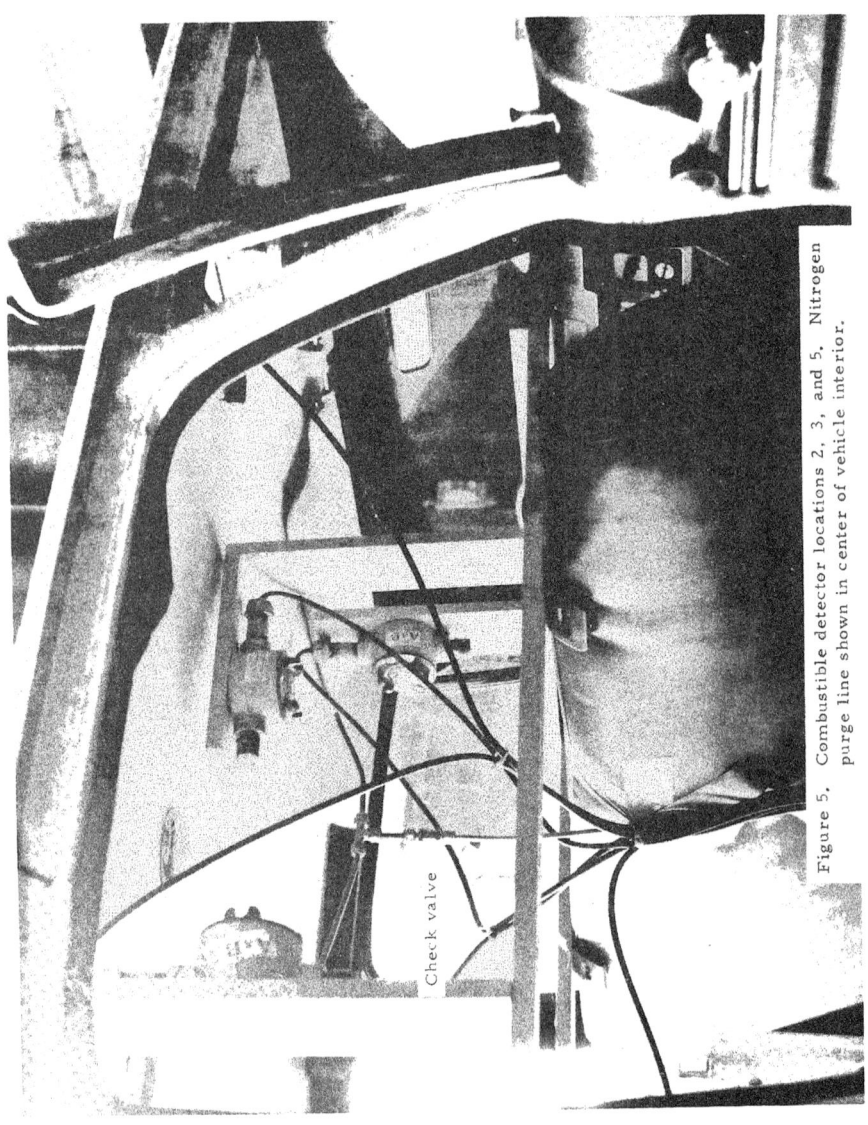

Check valve

Figure 5. Combustible detector locations 2, 3, and 5. Nitrogen purge line shown in center of vehicle interior.

14

with a water-filled U-tube manometer. No observable pressure differences could be detected at any time during the experiments with either $H_2$ or $CH_4$ gases. This result was anticipated as the chosen gas leakage rates were relatively low.

As indicated in figure 2, thermocouples were installed at sensor locations 5 to 8 to monitor the air-fuel temperature during a test. A permanent continuous record of temperatures was produced by connecting a strip chart recorder to the thermocouple outputs. The circuitry schematic for a typical sensor is shown in figure 4. These copper-constantan thermocouples were placed approximately one inch from the combustible gas detector heads, see figure 6. Reference junctions were maintained at 0°C and all splices in electrical leads, connecting thermocouples in the reference bath to instruments in the control room, were made in an insulated container designed for this purpose, see figure 7.

Leakage flow rates were measured using floating-ball flowmeters (see FM1 and FM2 on figure 2). Total volumetric flow (0.216 $m^3$ for all tests) was measured by using a calibrated commercial dry-gas meter, see figure 2.

Solid-state combustible gas sensors were used to detect gaseous fuel concentrations at various locations within the vehicle, see figure 2. Eight individual sensors were used, each consisting of a totally independent system. This type of gas analyzer measures the gas concentration at the sensing head and does not depend upon withdrawal of a sample to a remote analyzing instrument. Consequently, the gas analysis does not influence the mixture concentration in any way (other than the presence of the sensing head itself). Electrical outputs from all eight sensor leads were amplified and coupled to a multi-channel strip chart recorder.

The overall uncertainties of the instrumentation, including calibration and readout errors, etc., are as follows: Pressure, ± 1.65 mm $H_2O$;

15

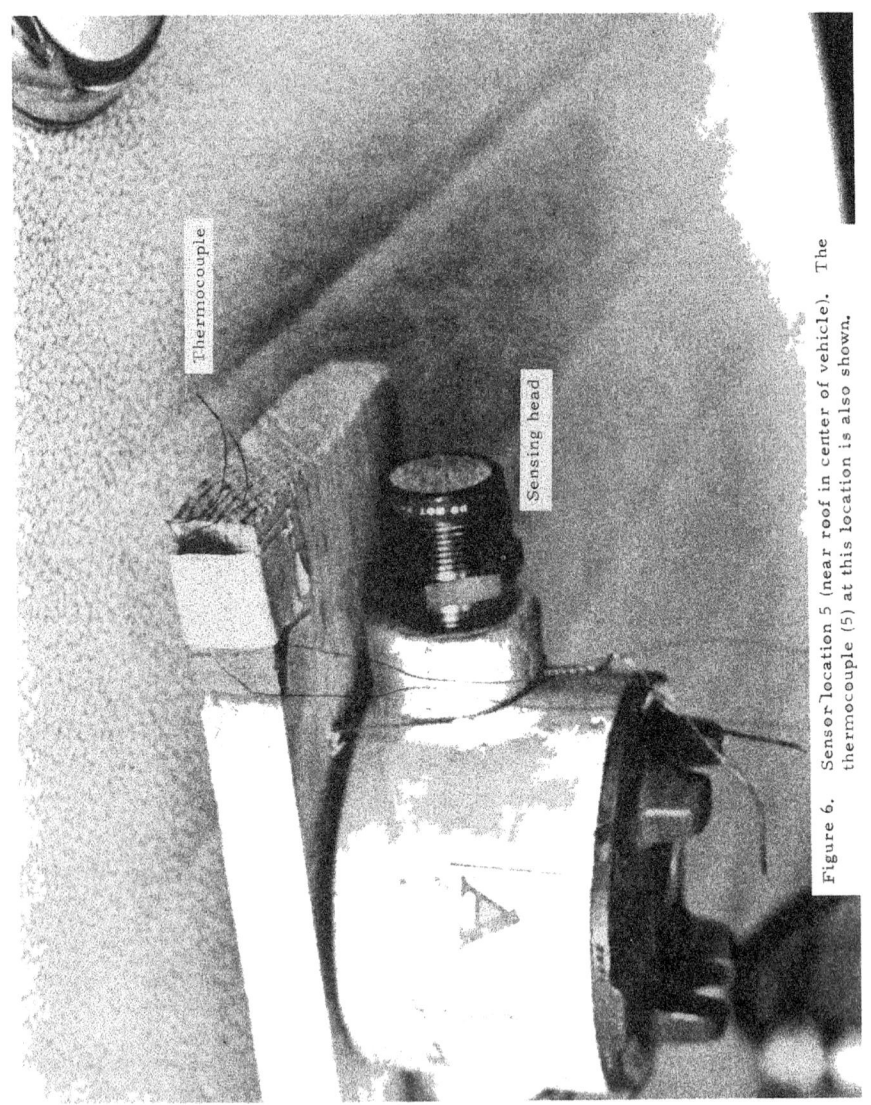

Figure 6. Sensor location 5 (near roof in center of vehicle). The thermocouple (5) at this location is also shown.

16

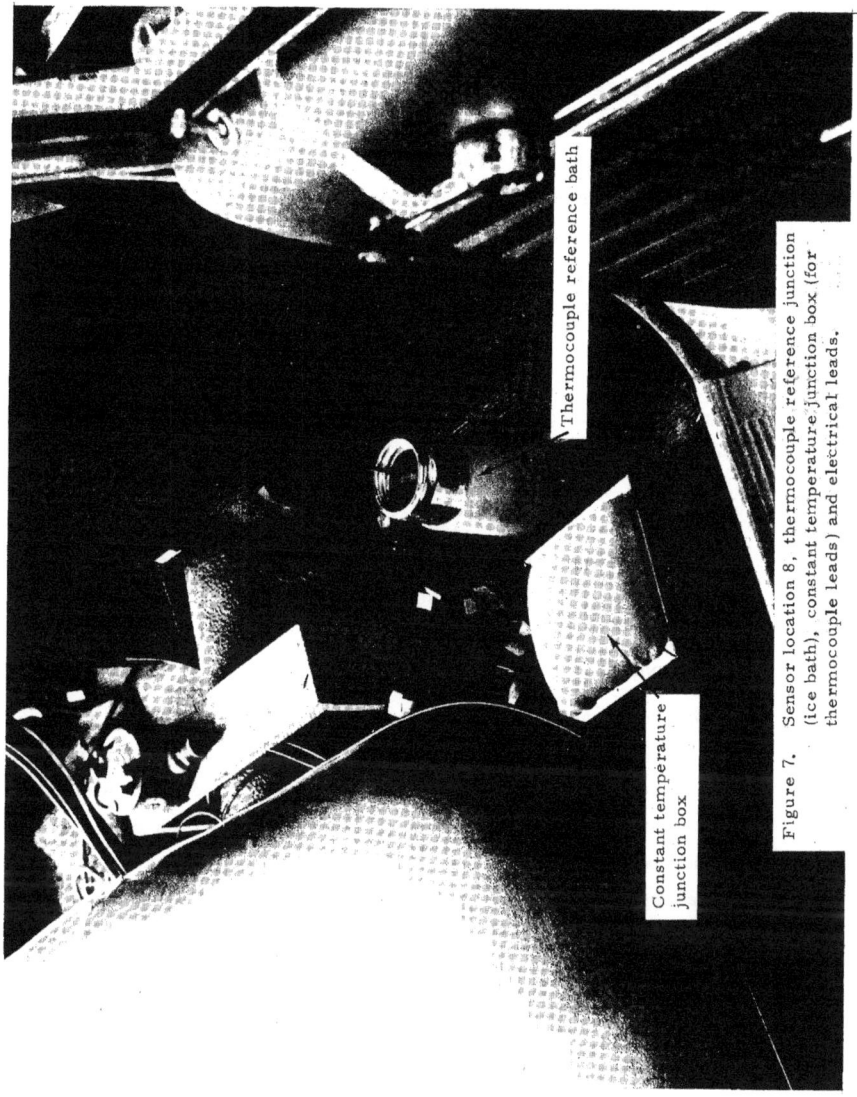

Figure 7.  Sensor location 8, thermocouple reference junction (ice bath), constant temperature junction box (for thermocouple leads) and electrical leads.

Thermocouple reference bath

Constant temperature junction box

17

temperature, $\pm$ 1 K; flow rate, $\pm$ 3 percent at all flow rates; total flow, $\pm$ 3 percent; gaseous fuel concentration in air, $\pm$ 3 percent of Lower Explosive Limit (LEL) for each fuel ($H_2$ or $CH_4$).

The combustible gas detectors were calibrated in-place many times during the test program. In the beginning of the program calibration checks were made before and after each test. This procedure was followed to assure that no change in calibration occurred during a test. As each detector (e.g., sensor No. 1) is a totally independent unit, all eight sensors were individually calibrated with identical gas mixtures. A known mixture of gas ($H_2$ or $CH_4$ in air) was used to pressurize an evacuated bladder through a seal-off clamp. Next, a special threaded fixture was attached to the end of the detector head located inside of the passenger compartment. The bladder was then attached to the fixture and the sensing head was ready for calibration. The combustible gas analyzer, located in the control room, was zeroed and the calibration mixture was released from the bladder. The analyzer unit then responded to the stagnant calibration mixture by indicating a concentration level. After detector stabilization was achieved, the analyzer instrument gain was adjusted so that the output indicator agreed with the calibration gas mixture exposed to the detector head.

For hydrogen detection 99.2 volume percent air and 0.8 percent $H_2$ (20% of the LEL) was used as the calibration gas mixture. Following the initial calibration, all hydrogen tests were conducted with periodic verifications of the initial calibration. For the series of runs involving methane, a mixture of 97.5 volume percent air and 2.5 percent $CH_4$ (47% of the LEL) was used to calibrate the sensing heads. Identical calibration procedures were used in both the $H_2$ and $CH_4$ tests.

## 5.0   TEST PROCEDURE

Tests were conducted with the auto non-vented and vented (three different roof and firewall vent areas were used). In both vented and non-vented tests all doors, windows and fresh-air vents were securely closed. Identical test procedures were used for both $H_2$ and $CH_4$ gases.

18

Prior to a test all electronic equipment, e.g., recorders, detectors, thermocouple amplifiers, etc., were warmed-up enough to assure reliable use. The test bay roof exhaust fan was turned on and the large access doors to the test bay were opened to expose the frangible plastic curtain. All system valves were secured in the closed position (see figure 4) and the tests performed as follows:

1. Zero combustible gas analyzer and record pre-test thermocouple readings (after commencement of test the thermocouple readings are taken at 5 to 10 minute intervals throughout the run.).

2. Start strip chart recorders.

3. Open V-6 (valve 6).

4. Open V-2 or V-3 (depending on test gas desired).

5. Regulate gas pressure to approximately 20 psig ($\sim$ 2.2 bar abs.)

6. Select flow rate and use appropriate flowmeter with its calibration curve. Throttle with V-4 or V-5.

| Test gas | FM1 Range (sccs) | FM2 Range (sccs) |
|---|---|---|
| $H_2$ | 0.0 to 657 | 0.0 to 103 |
| $CH_4$ | 0.0 to 251 | 0.0 to 42 |

7. After flow has stabilized initiate the run by simultaneously closing V-6 and opening V-7.

8. Monitor combustible gas sensor location No. 6 (located at gas inlet to car).

9. When a measurable concentration is detected at sensor No. 6 begin timing the run (response times ranged from $\sim$ 2 to 10 seconds for 36, 90, and 180 sccs flow rates).

10. Continuously monitor and record combustible levels at sensor locations 1 to 8 as a function of time.

11. When the required amount of combustible gas has been injected (0.216 $m^3$ for all tests) close V-7, open V-6, and turn off gas at bottle (close V-2 or V-3).

19

Table 1.   Summary of hydrogen dispersion data for a passenger vehicle.

| Figure | Flow Rate, sccs | Duration Gas Flow, Min. | Vents (cm²) | 1) During Gas Injection, Sensor Location | 2) @ Gas Cut-off, Sensor Location | 3) Decay Following Gas Cut-off, Sensor Location |
|---|---|---|---|---|---|---|
| 8 | 36 | 100 | 38.7 | 5 = 72% LEL @ 80 min.<br>1 = 63% " " " "<br>2 = 57% " " " "<br>3 = 53% " " " "<br>4 = 48% " " " "<br>6 = 41% " " " "<br>8 = 13% " " " "<br>7 = 2% " " " " | 1-8 = Approx. the same values as shown in 1) | 1-8 < 50% LEL @ 104 min.*<br>1-8 < 20% " " 122 ' |
| 9 | 36 | 100 | closed | 1-3,5 > LEL @ 28-33 min.<br>4 > " = 36 "<br>6 " : 42 " | 1-6 > LEL<br>7 = 53% LEL<br>8 = 48% " | 3 < LEL @ 106 min.<br>1,5 < " = " 118 "<br>2,4,6 < " = " 124 "<br>1-8 < 50% LEL @ 185 min.<br>1-8 < 20% LEL @ 345 min. |
| 10 | 90 | 40 | 38.7 | 5 > LEL @ 10.5 min. | 5 > LEL<br>2 = 86% LEL<br>1 = 83% "<br>3 = 79.5% "<br>4 = 70% "<br>6 = 61% "<br>8 = 11.5% "<br>7 = 2.5% " | 5 < LEL @ 41.5 min.<br>1-8 < 50% LEL @ 45 min.<br>1-8 ~ 10% " " 85 " |

* Total time from initiation of test.

Note:   sccs = standard cubic centimeters per second of gas flow (@20°C and 1 atm).
LEL = lower explosive limit based on volume percent of combustible gas in air.

20

Table 1 (continued).  Summary of hydrogen dispersion data for a passenger vehicle.

| Figure | Flow Rate, sccs | Duration Gas Flow Min. | Vents (cm$^2$) | 1) During Gas Injection, Sensor Location | 2) @ Gas Cut-off, Sensor Location | 3) Decay Following Gas Cut-off, Sensor Location |
|---|---|---|---|---|---|---|
| 11 | 90 | 40 | closed | 5 > LEL @ 4 min.<br>1-4 > " " 7-10 min.<br>6 > " " 14 min. | 1-6 > LEL<br>7 = 62% LEL<br>8 = 43% " | 3 < LEL @ 54 min.<br>1,2 < " " 68 "<br>4-6 < " " 82-84 min.<br>1-8 < 50% LEL @ 141 Min.<br>1-8 < 20% LEL @ 286 min. |
| 12 | 180 | 20 | 38.7 | 5 > LEL @ 2 min.<br>2 > " " 4 "<br>1,3 > " " 5 "<br>4 > " " 6 "<br>6 > " " 12 " | 1-6 > LEL<br>8 = 8% LEL<br>7 = 2% " | 1-6 < LEL @ 21-25 min.<br>1-8 < 50% LEL @ 30 min.<br>1-8 < 20% LEL @ 48 min. |
| 13 | 180 | 20 | closed | 1-5 > LEL @ 2-4 min.<br>6 > " " 5 " | 1-6 > LEL<br>7 = 37% LEL<br>8 = 30% " | 3 < LEL @ 44 min.<br>1,5 < " " 62 "<br>4 < " " 67 "<br>2,6 < " " 70 "<br>1-8 < 50% LEL @ 135 min.<br>1-8 < 20% " " 292 |

21

Table 2. Summary of methane dispersion data for a passenger vehicle.

| Figure | Flow Rate, sccs | Duration Gas Flow, Min. | Vents (cm²) | 1) During Gas Injection, Sensor Location | 2) @ Gas Cut-off, Sensor Location | 3) Decay Following Gas Cut-off, Sensor Location |
|---|---|---|---|---|---|---|
| 14 | 36 | 100 | 38.7 | 5 = 68% LEL @ 90 min.<br>1 = 57% " " "<br>3 = 48% " " "<br>2 = 46% " " "<br>4 = 41% " " "<br>6 = 24% " " "<br>8 = 9% " " "<br>7 = 1% " " " | 1-8 = Approx. the same values as shown in 1) | 1-8 < 50% LEL @ 104 min.<br>1-8 < 14% " 120 min. |
| 15 | 36 | 100 | closed | 5 > LEL @ 33 min.<br>1,4 > " " 35 "<br>2,3 > " " 40-41 min.<br>6 > " " 53 " | 1-6 > LEL<br>7 = 27% LEL<br>8 = 19% " | 2,3 < LEL @ 111 min.<br>5 < " " 113 "<br>1 < " " 115 "<br>4 < " " 117 "<br>6 < " " 118 "<br>1-8 < 50% LEL @ 171 min.<br>1-8 < 30% LEL @ 216 min. |
| 16 | 90 | 40 | 38.7 | 5 > LEL @ 13 min.<br>3 > " " 14 "<br>1 > " " 15 "<br>4 > " " 17 "<br>2 > " " 20 " | 1-5 > LEL<br>6 = 88% LEL<br>8 = 13% "<br>7 = 1% " | 2,3,4 < LEL @ 42 min.<br>1 < " " 43 "<br>5 < " " 47 "<br>1-8 < 50% LEL @ 52.5 min.<br>1-8 < 20% " 60 |

Note: sccs = standard cubic centimeters per second of gas flow (@ 20°C and 1 atm).

LEL = lower explosive limit based on volume percent of combustible gas in air.

22

Table 2 (continued).  Summary of methane dispersion data for a passenger vehicle.

| Figure | Flow Rate, sccs | Duration Gas Flow Min. | Vents (cm²) | 1) During Gas Injection, Sensor Location | 2) @ Gas Cut-off, Sensor Location | 3) Decay Following Gas Cut-off, Sensor Location |
|---|---|---|---|---|---|---|
| 17 | 90 | 40 | closed | 1,3,5 > LEL @ 9 min.<br>2 > " 10 "<br>4 > " 11 "<br>6 > " 13 " | 1-6 > LEL<br>7 = 43% LEL<br>8 = 36% " | 6 < LEL @ 66 min.<br>4 < " 72 "<br>2 < " 72.5 min.<br>3 < " 76 "<br>5 < " 84 "<br>1 < " 86 "<br><br>1-8 < 50% LEL @ 133 min.<br>1-8 < 33% " " 200 |
| 18 | 180 | 20 | 38.7 | 1,2,5 > LEL @ 2.5 min.<br>3,4 > " 3 "<br>6 > " 5 " | 1-6 > LEL<br>8 = 5% LEL<br>7 = ~0.5% LEL | 6 < LEL @ 23 min.<br>2,3,4 < " 26 "<br>1 < " 27 "<br>5 < " 32 "<br><br>1-8 < 50% LEL @ 39.5 min.<br>1-8 < 20% " " 50.5 |
| 19 | 180 | 20 | closed | 1-5 > LEL @ 3-4 min.<br>6 > " 5 " | 1-6 > LEL<br>7 = 17% LEL<br>8 = 13% " | 2 < LEL @ 85 min.<br>1 < " 87 "<br>3 < " 88 "<br>4 < " 91 "<br>5 < " 95 "<br>6 < " 98 "<br><br>1-8 < 50% LEL @ 158 min.<br>1-8 < 30% " " 202 |

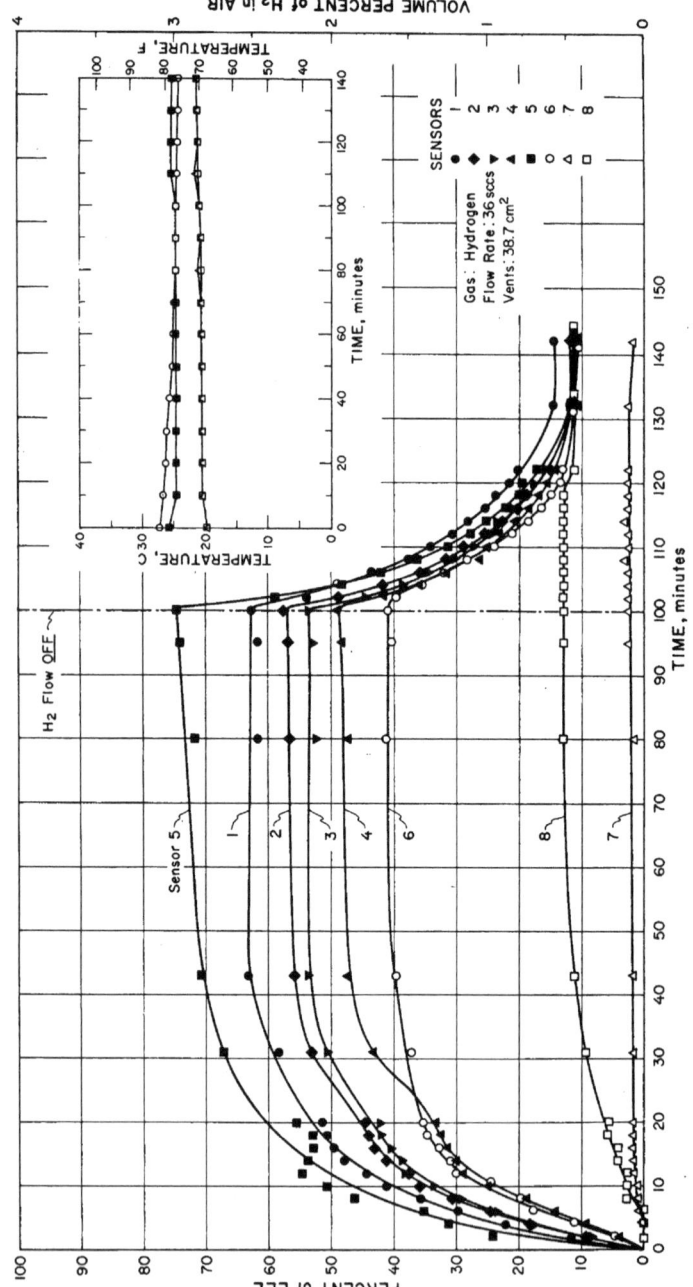

Figure 8. Dispersion characteristics of $H_2$ gas in the passenger compartment of a 1970 sedan.

24

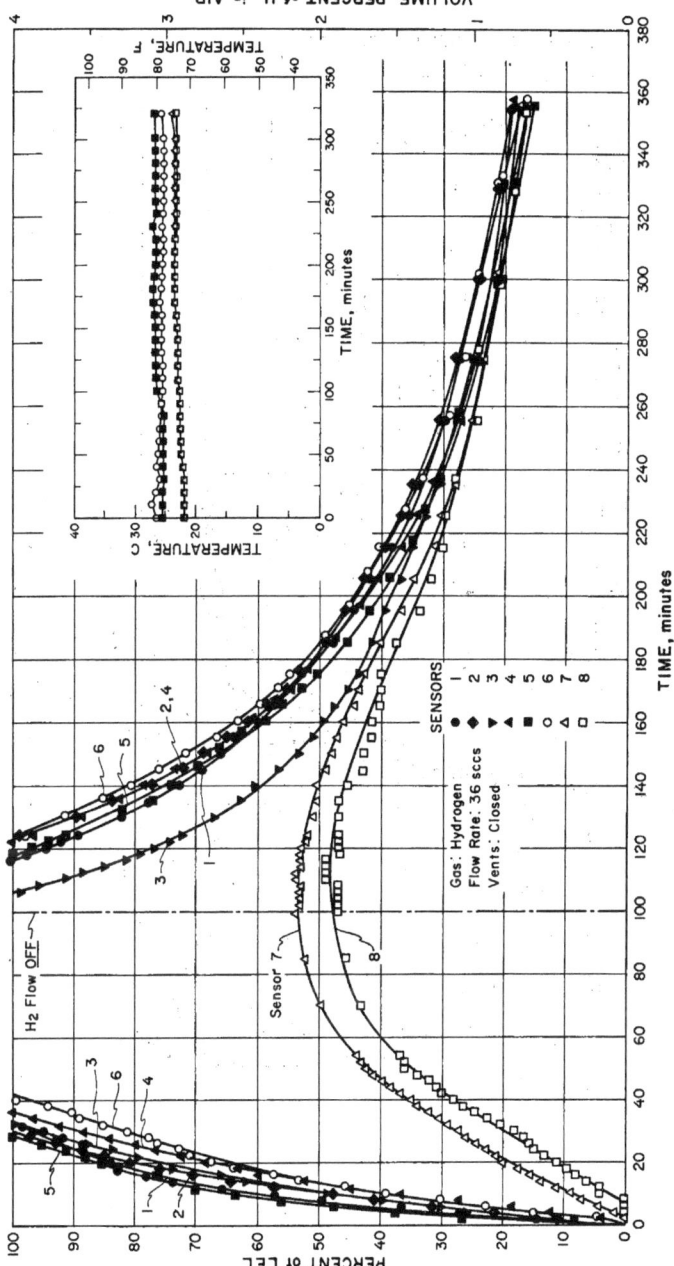

Figure 9. Dispersion characteristics of $H_2$ gas in the passenger compartment of a 1970 sedan.

25

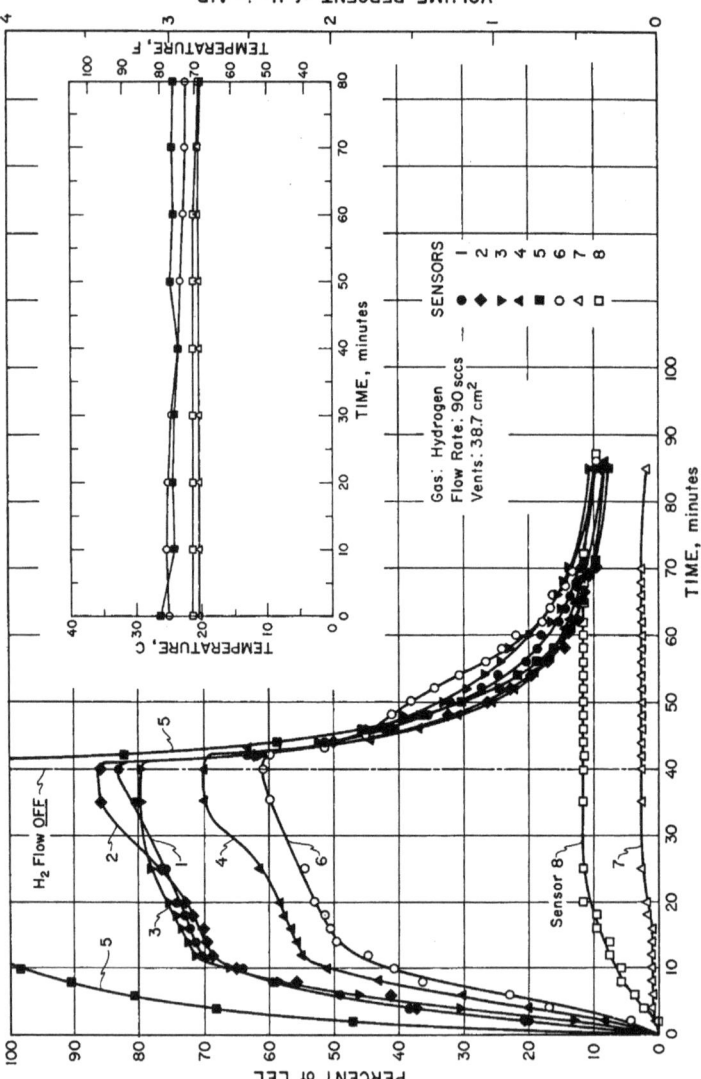

Figure 10. Dispersion characteristics of H₂ gas in the passenger compartment of a 1970 sedan.

26

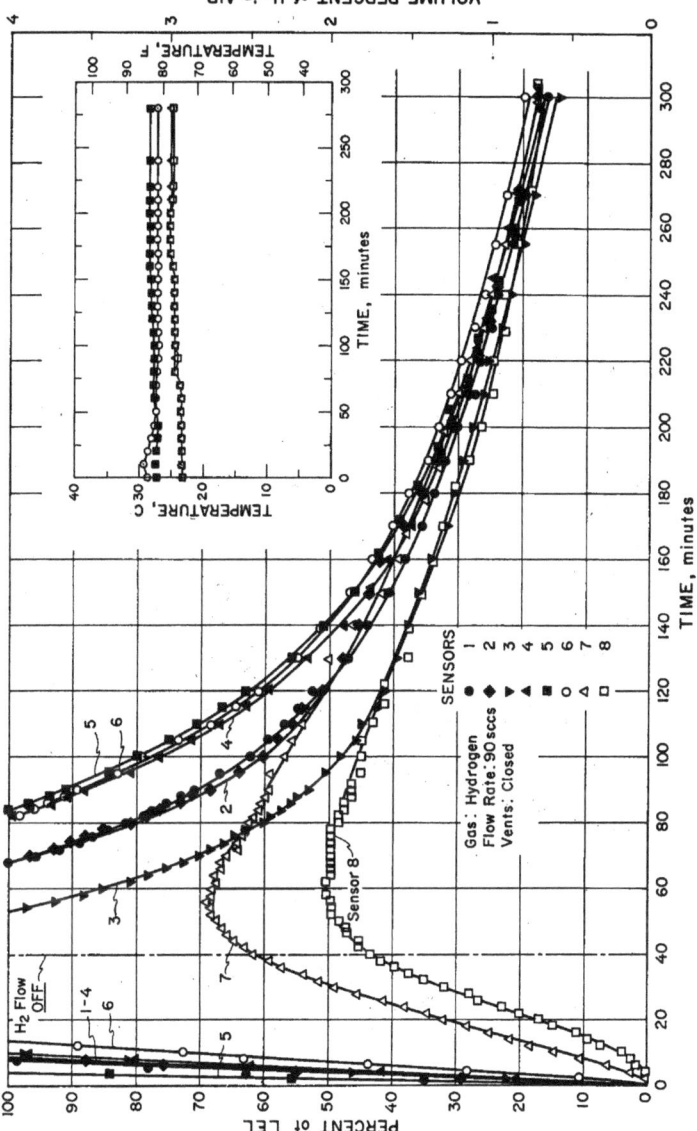

Figure 11. Dispersion characteristics of $H_2$ gas in the passenger compartment of a 1970 sedan.

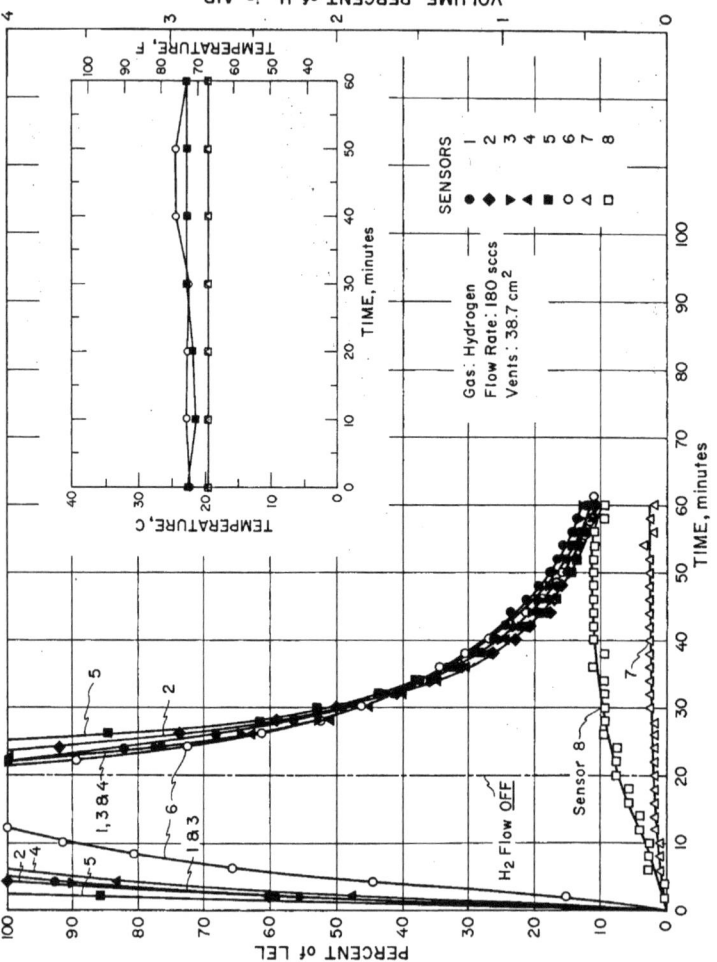

Figure 12. Dispersion characteristics of $H_2$ gas in the passenger compartment of a 1970 sedan.

28

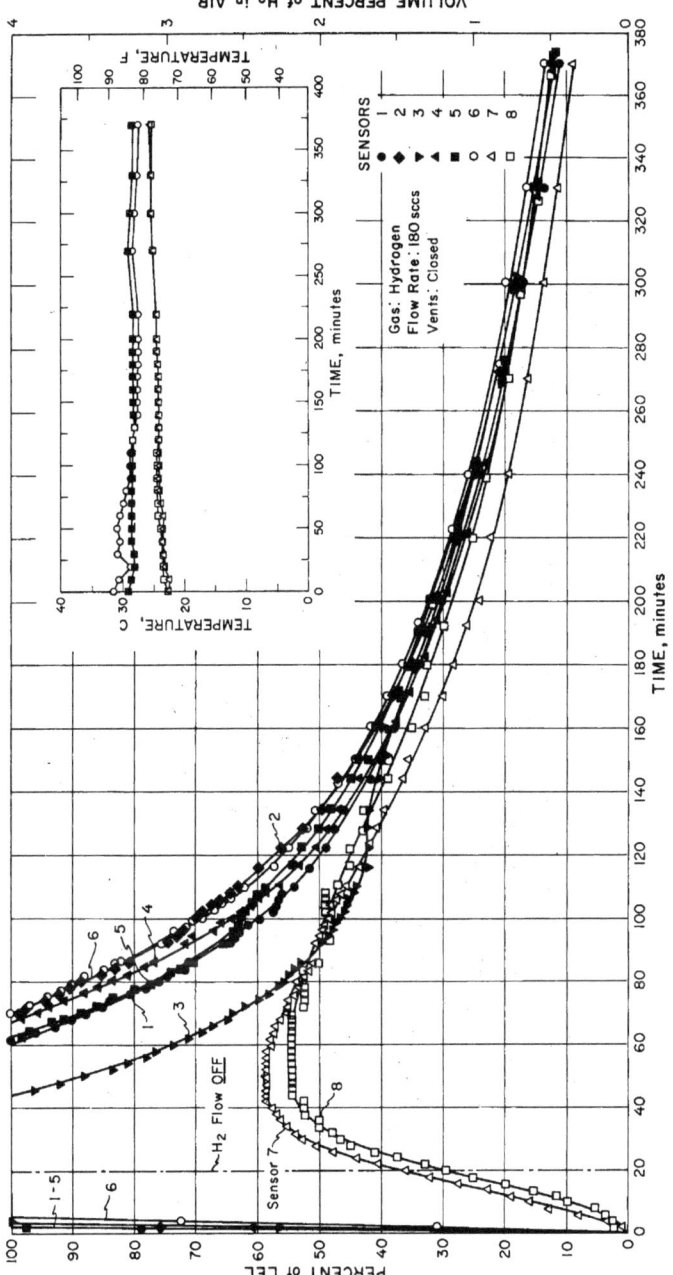

Figure 13. Dispersion characteristics of H$_2$ gas in the passenger compartment of a 1970 sedan.

29

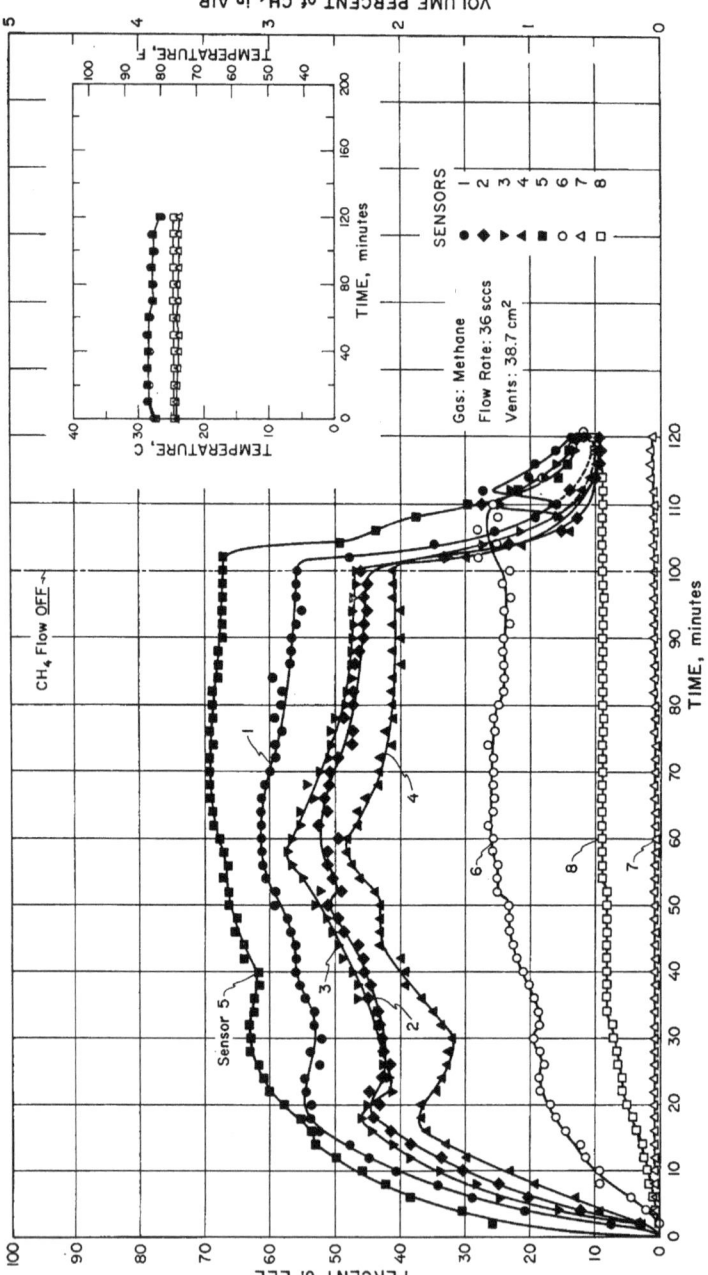

Figure 14. Dispersion characteristics of $CH_4$ gas in the passenger compartment of a 1970 sedan.

30

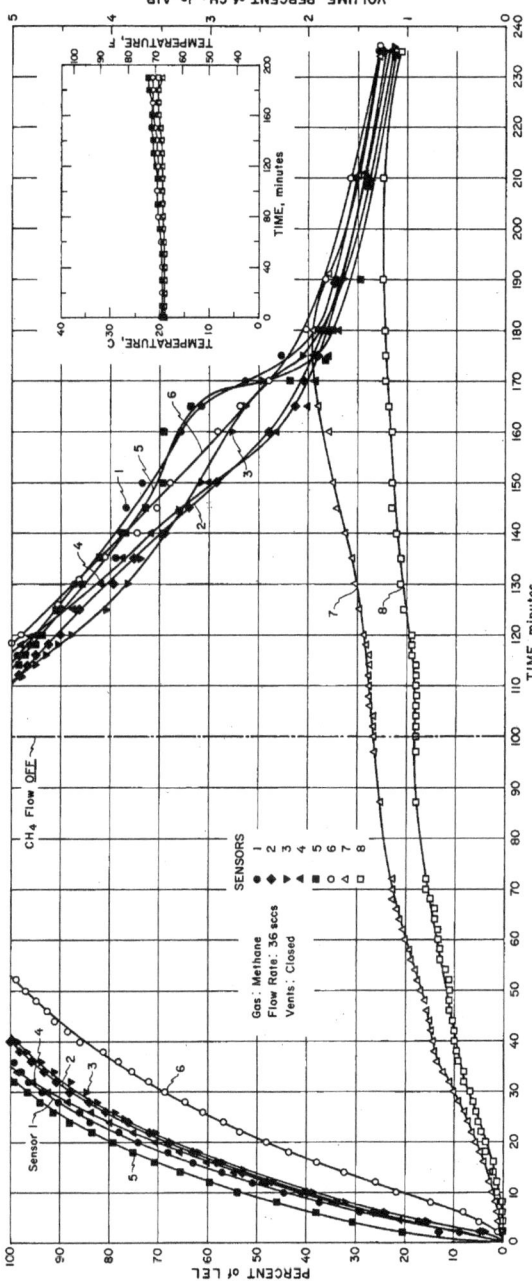

Figure 15. Dispersion characteristics of CH₄ gas in the passenger compartment of a 1970 sedan.

31

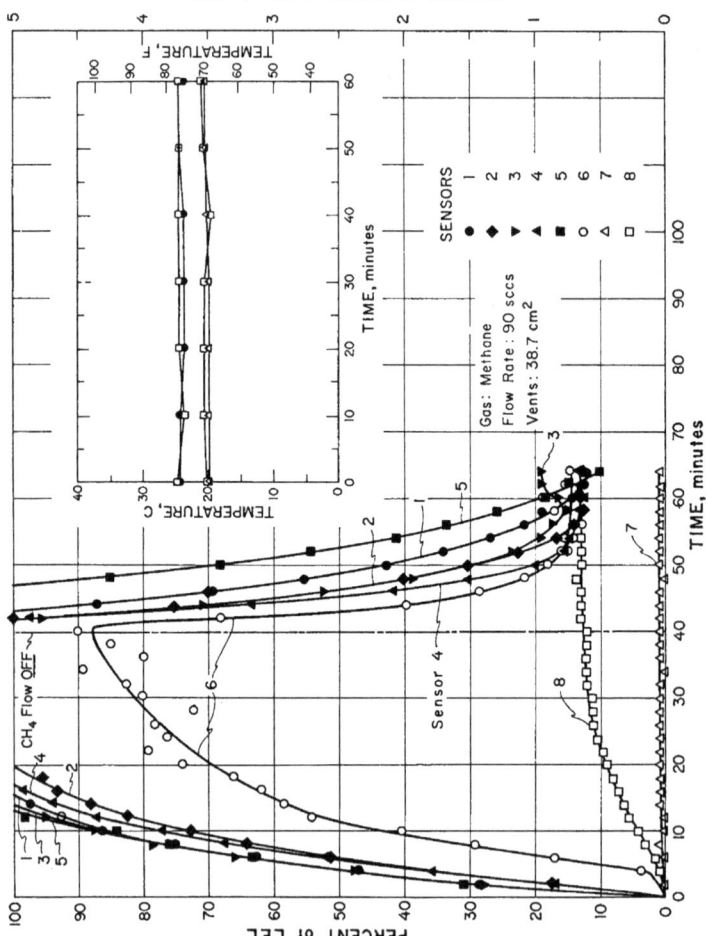

Figure 16. Dispersion characteristics of CH₄ gas in the passenger compartment of a 1970 sedan.

32

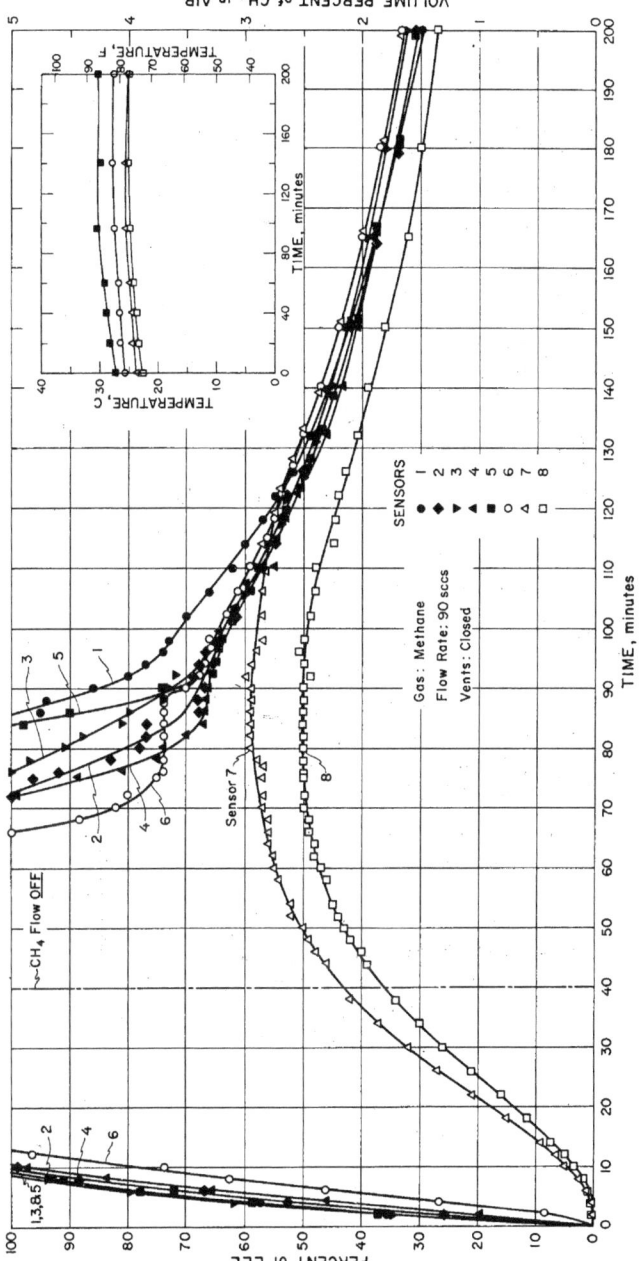

Figure 17. Dispersion characteristics of CH₄ gas in the passenger compartment of a 1970 sedan.

33

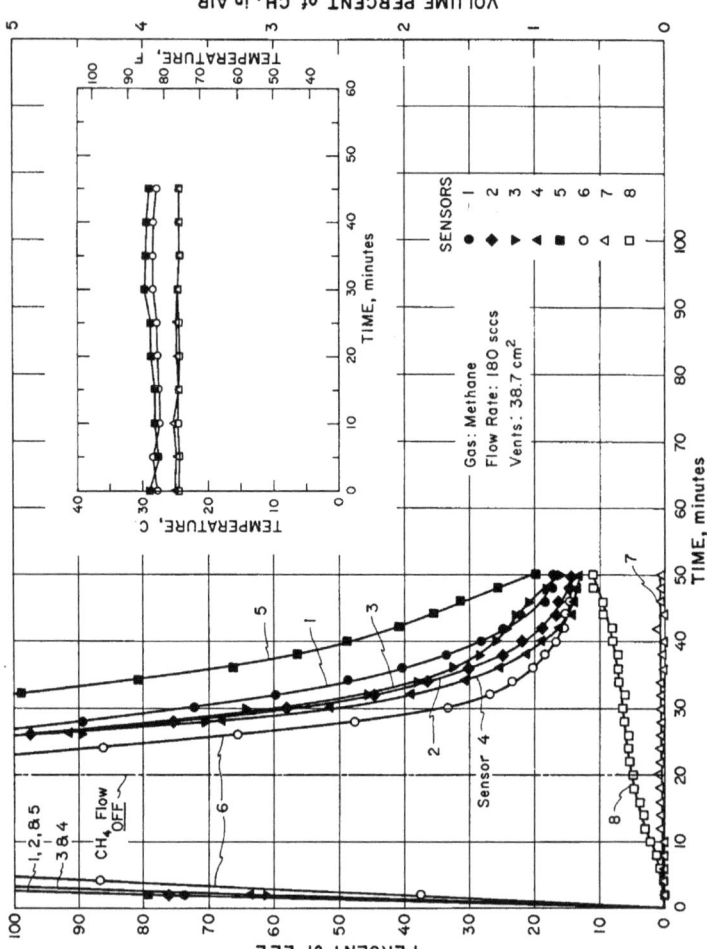

Figure 18. Dispersion characteristics of CH$_4$ gas in the passenger compartment of a 1970 sedan.

34

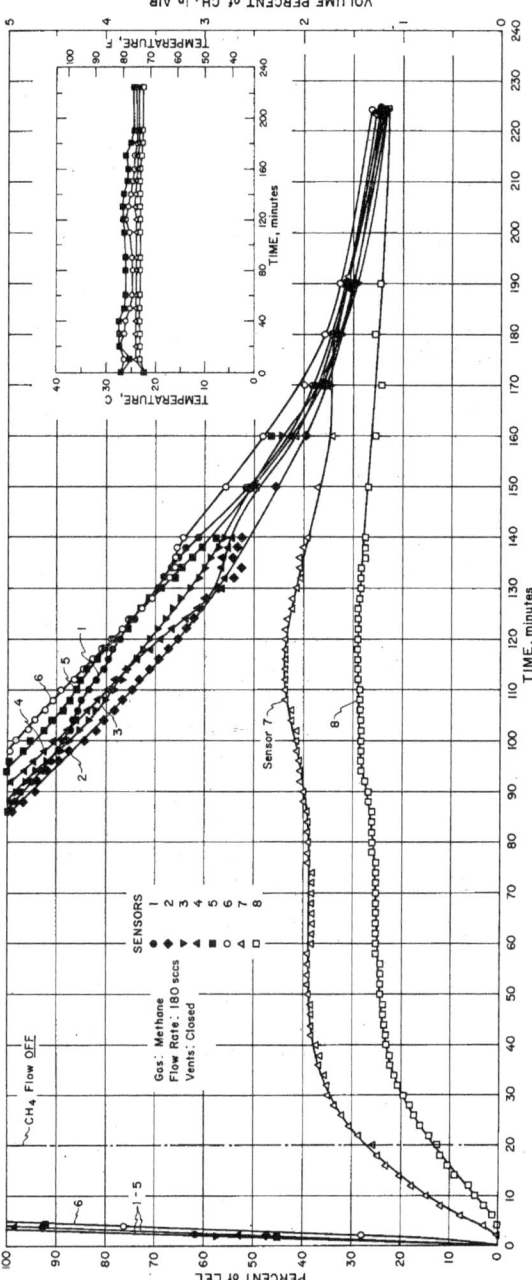

Figure 19. Dispersion characteristics of $CH_4$ gas in the passenger compartment of a 1970 sedan.

35

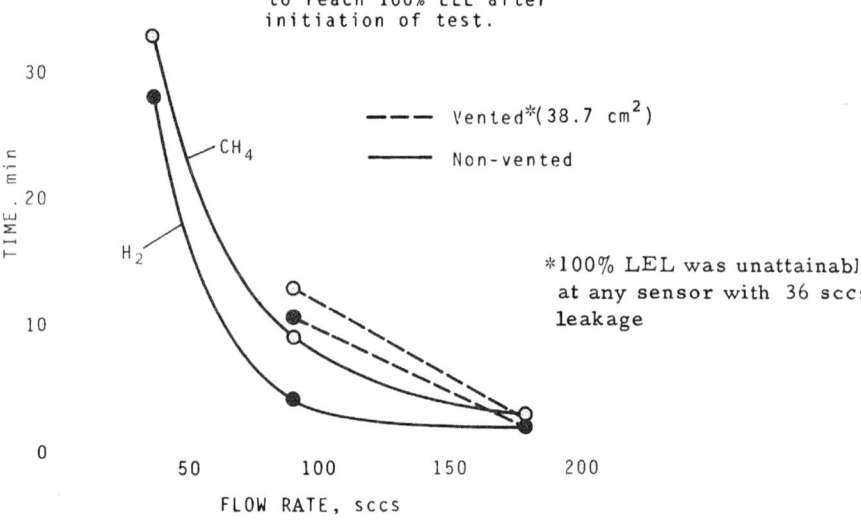

to reach 100% LEL after
initiation of test.

- - - - Vented*(38.7 cm$^2$)
———— Non-vented

*100% LEL was unattainable
at any sensor with 36 sccs
leakage

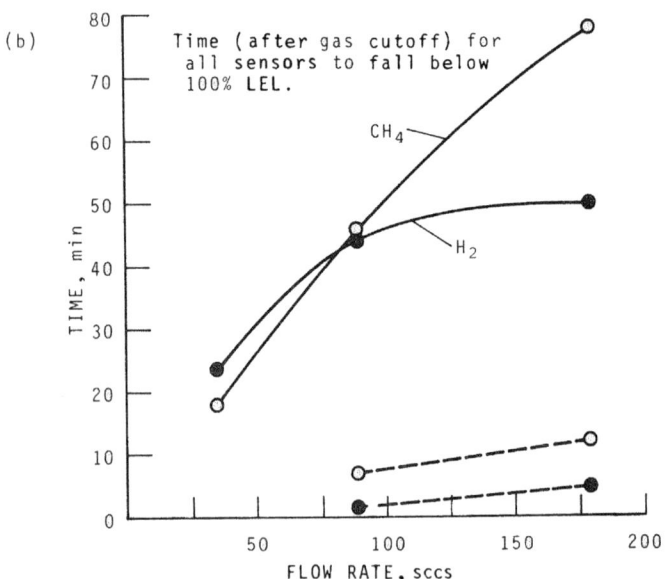

(b)

Time (after gas cutoff) for
all sensors to fall below
100% LEL.

Figure 20. Threshold concentration (100% LEL) as a function of time
and flow rate for $H_2$ and $CH_4$ gases in the passenger
compartment of a 1970 sedan.

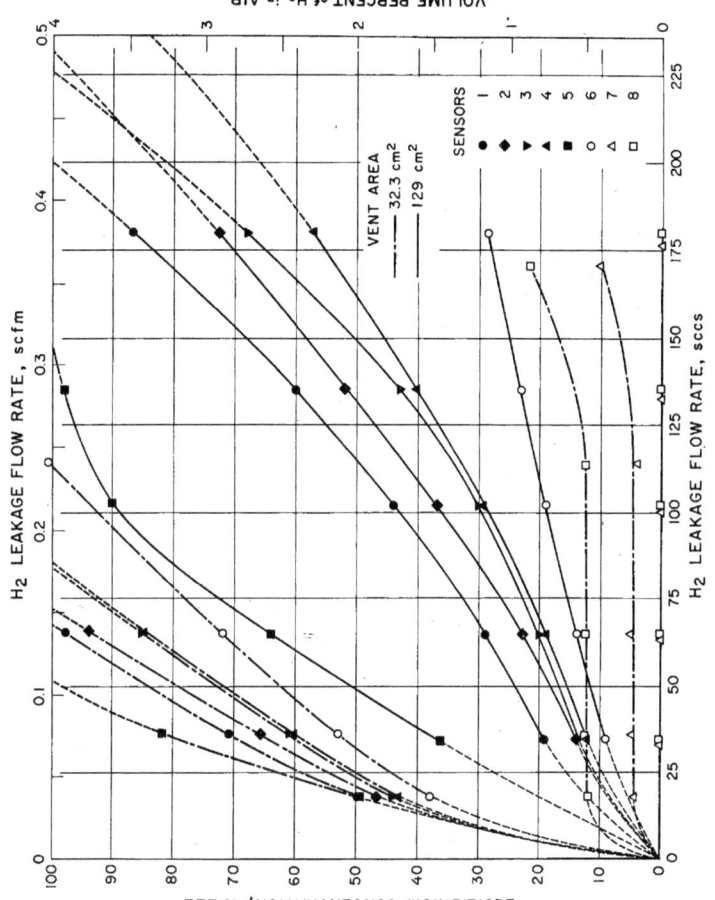

Figure 21. Effect of vent area and leakage flow rate on equilibrium concentrations of $H_2$ gas in the passenger compartment of a 1970 sedan.

37

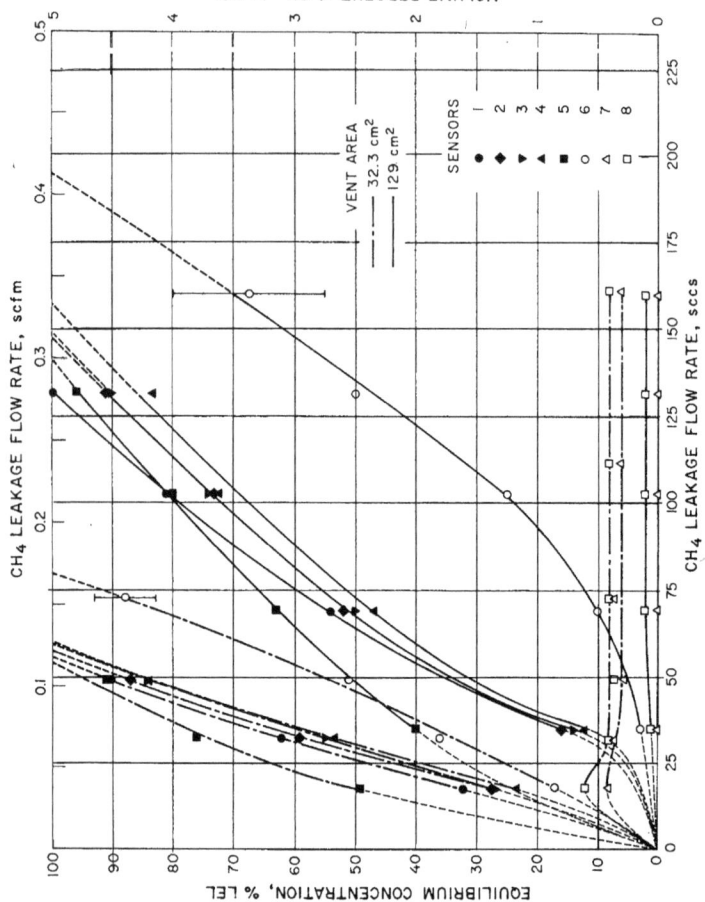

Figure 22. Effect of vent area and leakage flow rate on equilibrium concentrations of CH₄ gas in the passenger compartment of a 1970 sedan.

38

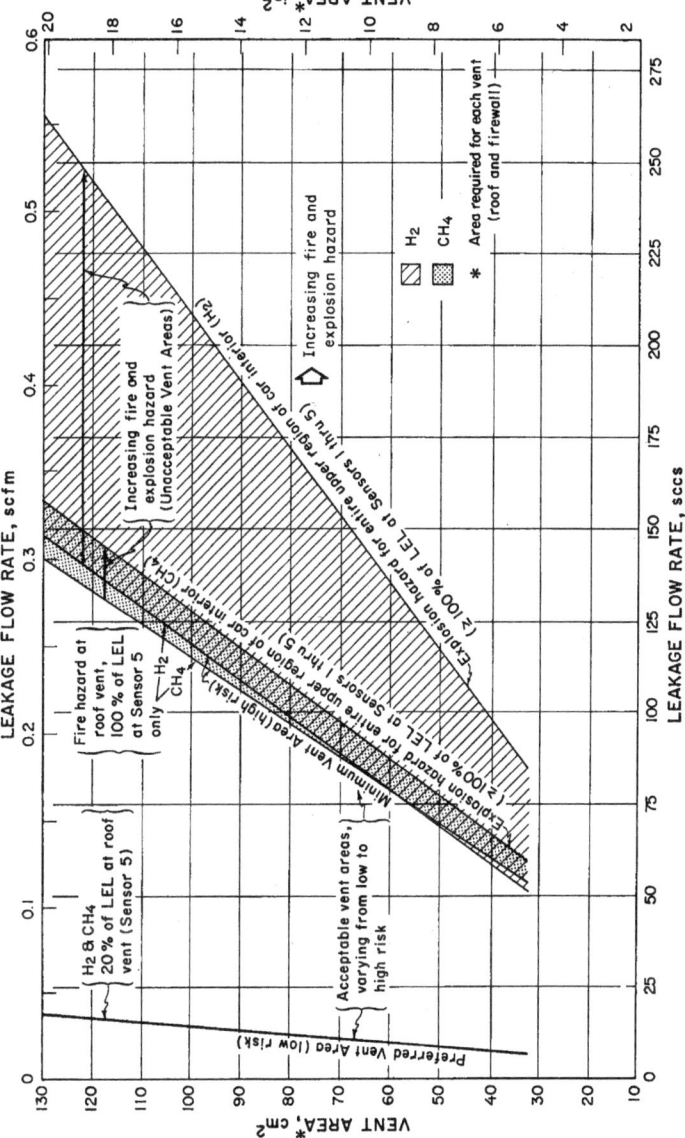

Figure 23. Potential fire and explosion hazards in the passenger compartment of a 1970 sedan as a function of vent area and leakage flow rate.

12. Continue to monitor and record the decay of concentration levels
    for all detectors down to approximately 20 to 30 percent of LEL.
13. Terminate the run by purging the vehicle interior with nitrogen.
    Open V-1, close V-6, and open V-8 -- the $N_2$ bottle pressure
    regulator is set at approximately 25 psig ($\sim$ 2.5 bar abs.).
14. Allow sufficient flow of $N_2$ into the vehicle for adequate dilution
    of the combustible gas.
15. Open the car doors to achieve thorough $N_2$ purge and ventilation.
16. Close V-1 and V-8.

With the building-controlled environment and the test procedure
outlined above test data were highly reproducible (not demonstrated
herein but verified by several repeat tests at identical flow rates and
vented-car conditions).

## 6.0  TEST RESULTS

The test data are summarized in tables 1 and 2 and are shown gra-
phically in figures 8 to 23.  The combustible sensor data designated on
figures 8 to 19 correspond to those locations schematically illustrated
in figure 2.  Temperature vs time plots for each test are shown on each
of figures 8 to 19 and the thermocouple locations (5 to 8) are coincident
with the mixture sensor locations.

Figures 8 to 13 illustrate the dispersion characteristics of hydrogen
at three flow rates and with two venting configurations (non-vented and
with 38.7 $cm^2$ vents).  The methane test results shown in figures 14 to 19
are at the same flow rates and venting conditions stated above for
hydrogen.  This allows a direct comparison of hydrogen and methane dispersion data.  Figures 9, 11, and 13 have an abscissa incremented in
20 minute intervals while all other figures have 10 minute increments.

The equilibrium concentration (percent of LEL) is given as a function
of leakage flow rate and vent area in figures 21 and 22 for $H_2$ and $CH_4$.
From these data relationships between vent area, leakage flowrate and
equilibrium gas mixtures for $H_2$ and $CH_4$ may be derived.  One such

40

relationship is illustrated on figure 23. The equilibrium concentration as designated on figures 21 to 23 is the steady-state gas mixture achieved under sustained leakage flow.

Figure 20a shows the time required for any one of the eight sensors to reach 100 percent of LEL after initiation of the test (vented or non-vented); figure 20b gives the time required (after leakage flow cutoff) for all sensors to fall below 100 percent of LEL.

## 7.0 DISCUSSION OF TEST RESULTS

The data shown on figures 8 to 19 are considered representative of the dispersion behavior of hydrogen and methane when released inside of a passenger vehicle. These figures illustrate how quickly the LEL is reached and in some cases how long combustible gas concentrations can remain inside the vehicle. Several repetitive tests with each gas exhibited remarkable reproducibility and strengthened our confidence in the test results. While it is recognized that these data are precisely applicable to but one auto, the test vehicle, it is felt that the results are generally applicable to all passenger vehicles.

Some of the more important trends obviated by careful study of figures 8 to 19 are: (1) vents in the roof and firewall (38.7 $cm^2$ each) effectively delay explosive accumulations of $H_2$ or $CH_4$; (2) vents also promote more rapid dilution of explosive gas mixtures ($H_2$-air or $CH_4$-air) after leakage flow is stopped; (3) the lowest leakage flows (probably most typical of loose fittings, cracked welds, tubing splits, trunk membrane leaks, etc.) do not produce combustible gas mixtures in tests with 38.7 $cm^2$ vents, see figures 8 and 14; (4) the best location for an early-warning combustible gas detector is at the top-center of the vehicle near the roof vent (sensor 5 in these tests).

These tests show that non-vented vehicles may retain hazardous gas mixtures for an hour or more -- about an order of magnitude longer than for vented vehicles. This result suggests that a ventilation system

41

--forced or natural (as in these tests) -- is very desirable for vehicles
converted to operate on gaseous fuels. Vehicles designed to operate on
gaseous fuels could perhaps expunge this ventilation criterion by
proper placement of a well-vented fuel system exterior to a well-sealed
vehicle. Even so early-warning explosive gas detectors may be well advised.

To compare $H_2$ and $CH_4$ fuels we must compare the test results from
identical tests, e.g., pair figures 8 and 14, 9 and 15, etc. At the
lowest flow rate we observe similar results for the two gases whether
the vehicle is vented or closed (figures 8 and 14, 9 and 15). $H_2$
disperses more rapidly than $CH_4$ with intermediate leakage into a vented
vehicle (figures 10 and 16); however, the two gases behave similarly with
intermediate leakage flow into a closed vehicle (figures 11 and 17).
At the highest flow rate $H_2$ disperses more rapidly than $CH_4$ in both the
vented and closed vehicles (figures 12 and 18, 13 and 19). These ob-
servations support the contention that buoyant flow mechanisms are
predominant and dispersion by diffusion is nearly negligible -- refer to
rationale in section 3.0.

A generalized summary of figures 8 to 19 is shown on figure 20.
A threshold gas concentration of 100 percent of LEL at any location
within the car is used to illustrate mixing and efflux rates for the
two gases. Higher leakage flow rates create combustible concentrations
very rapidly in both the vented and non-vented vehicle configurations,
see figure 20a. Figure 20b clearly illustrates the advantages of
vehicle vents. Again, it is not surprising that $H_2$ reaches, and
decays below, LEL concentrations more rapidly than $CH_4$. Buoyant forces
favor the rapid dispersion of hydrogen over methane by the ratio of
1.45:1 and diffusion velocities favor $H_2$ by the ratio of 3:1.

Tests were also performed with vent areas (roof and firewall) of
32.3 and 129 $cm^2$. These tests were designed to evaluate the effect
of vehicle vent area on mixing and efflux rates at various leakage flow
rates. The time parameter was eliminated from these tests by maintaining

42

leakage flow steady until all eight gas sensor outputs were constant, i.e., an equilibrium (quasi steady-state) gas concentration was established at each sensor location in the passenger compartment. Thus it is possible to determine the vent area required, at each leakage flow rate, to avoid combustible gas mixtures at any location inside of the vehicle.

Such data, for $H_2$ and $CH_4$, are shown in figures 21 to 23. Larger vent areas accomodate larger leakage flow rates without reaching 100 percent of the LEL as illustrated in figures 21 and 22. Sensor 5, at the roof vent, exhibited the highest concentration of combustible gas in all of our tests. Consequently, data from sensor 5 as shown in figures 21 and 22 were used to derive vent area requirements for vehicles subjected to potential $H_2$ or $CH_4$ leaks. These vent requirements are plotted for two separate safety criteria on figure 23. The 20 percent of LEL criterion was selected because it is the industrially accepted limit for warning personnel of explosive hazards. The 100 percent of LEL criterion provides a true indication of incipient fire and explosion hazard.

Figure 23 shows that identical $H_2$ and $CH_4$ leakage flows require the same vent areas to maintain combustible gas concentrations below 20 percent of the LEL. Also, the incipient fire hazard (100 percent of LEL at sensor 5 near the roof vent) is almost identical for $H_2$ and $CH_4$ at the same leakage flows and vent areas. These leftmost boundaries of the shaded (high risk) regions on figure 23 are considered mainly as a fire hazard because only a small pocket of combustible gas exists near the roof vent. With the same vent area and higher flow rates, additional gas sensors reach the LEL and the fire and explosion hazards are greatly amplified. The width of the shaded regions on this figure indicates the tolerance of a vented vehicle to increased leakage flow of $H_2$ or $CH_4$. The narrow bandwidth for $CH_4$ means that slightly increased leakage flows in a specific vehicle result in explosive gas mixtures throughout the upper region of the passenger compartment. By comparison,

43

$H_2$ leakage must be significantly increased to fill the entire upper region of the car interior with a flammable $H_2$-air mixture. An example, explaining the use of figure 23 is given in Appendix B.

Figure 23 was composed from test data acquired with two vent areas (32.3 and 129 $cm^2$); therefore, the straight-line boundaries shown may exhibit some slight curvature with the benefit of additional test data. It was felt that additional tests, on this specific vehicle under the prescribed test conditions, were unnecessary.

## 8.0  CONCLUSIONS

$H_2$ and $CH_4$ gases were deliberately leaked into the passenger compartment of a 1970 sedan. Tests were conducted with the car doors and windows closed and in vented and non-vented configurations. Leakage flow rates ranged from 19 to 180 sccs and vehicle vent areas varied from 0.0 to 129 $cm^2$. Combustible gas sensors, placed inside of the vehicle, were used to detect mixing and efflux characteristics of the combustible gases. The test results reported here are specifically applicable to but one vehicle; however, the results are believed to be characteristic and generally applicable to all passenger vehicles.

1.  Vents in the roof and firewall effectively delay, and for low flows prevent, explosive accumulations of $H_2$ or $CH_4$ in the passenger compartment. Low leakage flows considered typical of loose fittings, cracked welds, membrane leaks from well sealed and vented trunks, etc., do not produce flammable gas mixtures in tests where the vent area $\geq$ 38.7 $cm^2$.

2.  Vents promote more rapid dilution of combustible gas mixtures after leakage flow is stopped. Non-vented vehicles may hold flammable mixtures for over an hour -- nearly 10X the holding times for vented vehicles.

3.  High leakage flows of $H_2$ or $CH_4$ produce combustible concentrations very rapidly in both vented and non-vented vehicles. $H_2$ reaches and decays below flammable concentrations more rapidly than $CH_4$.

4. For identical leakage flow rates of $H_2$ or $CH_4$, identical vent areas are required to avoid an incipient fire hazard in the car interior (even though $H_2$ disperses more rapidly). Criteria for acceptable vent areas — with varying levels of risk -- are defined.

5. Buoyant flow mechanisms (rather than diffusion) were determined to be the dominant mode of gaseous dispersion.

6. The prime location for an early-warning combustible gas detector is at the topmost point of the vehicle roof, near the roof vent.

7. On the basis of these tests we conclude that neither $H_2$ nor $CH_4$ gas is safer than the other as a vehicular fuel and that the two gases appear equally safe if used in properly designed vehicles. These experimental results indicate that progressive development of $H_2$ and $CH_4$ fueled autos can be actively pursued with the full expectation that such autos (properly engineered conversions and new designs) can ultimately gain the acceptance of regulatory organizations and the general public.

## 9.0  RECOMMENDATIONS

This study indicates that gasoline-powered vehicles converted to burn gaseous fuels should be equipped as follows: (1) The fuel system should be adequately designed for fuel containment in the event of collision; (2) the car trunk should be adequately vented if the fuel tank is located in the trunk; (3) the fuel tank vent should be ducted to the far rear exterior of the vehicle; (4) and a trunk membrane should be provided to isolate the trunk (or fuel) and passenger compartments. Optional, but highly desirable equipment for such vehicles includes a positive ventilation system and an early-warning combustible gas detector. Guidelines for the design of appropriate ventilation systems are given herein. If used, a combustible gas sensor should be located in the passenger compartment at the highest point of the roof (vented or non-vented).

It is recognized that auto heating and air conditioning systems are not capable of handling the increased heating and cooling loads imposed by the large vents prescribed in this paper. A more practical solution is to locate the fuel supply system, as with conventional gasoline-powered vehicles, on the exterior of the vehicle. The fuel system ($H_2$ or $CH_4$) should be an integral part of the vehicle design and located so that it can be well vented, readily serviced, protected from foul weather and collision, and isolated from the passenger compartment. With this integral design feature it seems plausible to omit the roof and firewall vents; however, until we gain more experience with $H_2$ and $CH_4$ fueled cars it also seems advisable to install a combustible gas warning system.

## 10.0 REFERENCES

1. Billings, R., Hydrogens potential as an automotive fuel, Cryogenics and Industrial Gases, Vol. 9, No. 1, 23-5 (1974).

2. Billings, R. E., and F. E. Lynch, History of hydrogen-fueled internal combustion engines, Billings Energy Research Corp., Publication No. 73001 (1972).

3. Weil, K. H., The hydrogen I.C. engine- its origins and future in the emerging energy-transportation-environment system, Proceed. of the 7th IECEC, 1355-1363 (1972).

4. Murray, R. G., R. J. Schoeppel, and C. L. Gray, The hydrogen engine in perspective, Proceed. of the 7th IECEC, 1375-81 (1972).

5. Swain, M. R., and R. R. Adt Jr., The hydrogen-air fueled automobile, Proceed. of the 7th IECEC, 1382-87 (1972).

6. Corbett, P. F., Natural gas and the environment - after combustion, Outlook for Natural Gas - A Quality Fuel, from the Proceedings of the Institute of Petroleum Summer Meeting, held at the Palace Court Hotel, Bournemouth, England, p. 221-235 (6-9 June 1972).

7. Anon., Emission reduction using gaseous fuels for vehicular propulsion, Final report on contract no. 70-69, Institute of Gas Technology, IIT Center, Chicago, Ill. 60616, (June 1971).: Also PB201 410, Available from NTIS.

8. Johnson, E. F., Fire protection developments in CNG-fueled vehicle operations, Fire Journal Vol. 66, No. 6 (Nov. 1972).

9. Anon., Pollution reduction with cost savings (report on the General Services Administration's dual-fuel experiment), GSA DC71-10828 (1971).

10. Shooter, D. and A. Kalelkar, The benefits and risks associated with gaseous fueled vehicles, report to the Massachusetts Turnpike Authority, Arthur D. Little Case 74400-2 (May 1972).

11. Van Vorst, W. D., (Univ. of Calif., Los Angeles, Calif.) private communication to the authors; and Bush, A. F., and Van Vorst, W. D., The UCLA hydrogen car, Book, Advances in Cryogenic Engineering 19, Ed. K. D. Timmerhaus, pp. 23-27 (Plenum Press, Inc., New York, N.Y., 1974).

12. Enserink, E., Dual-fuel motor vehicle safety impact testing, DoT/HS-800 622, Available from NTIS, (Nov. 1971).

13. Docket No. HM-115; Notice No. 74-3, Transportation 49 CFR Parts 172, 173, 177, 178, 179, Cryogenic Liquids, U.S. Department of Transportation, Office of Hazardous Materials, 2100 - 2nd St., S.W., Washington, D.C. 20590, provided by A. J. Mallen in private communication to the authors.

14. Ryan, L. E., Massachusetts Turnpike Authority, Suite 3000, Prudential Center, Boston, Mass. 02199, private communication to the authors.

15. Proposed revisions to California regulations for LNG motor vehicle fuel containers and regulations for compressed and liquefied gas fuel systems, California Administrative Code, Title 13, Chapter 2, Article 2, Sections 930-936, Department of California Highway Patrol, P.O. Box 898, Sacramento, Calif. 95804: provided by W. M. Heath in private communication to the authors. Also the Compressed Gas Association, 500 Fifth Ave., New York, N.Y. 10036, is cooperating with the State of California and DoT to develop specifications for the design and construction of LNG motor fuel containers.

16. An editorial staff summary of California regulations for LNG auto installations, Cryo. Tech., 132-134 (July-Aug. 1972).

17. Hord, J., Correlations for predicting leakage through closed valves, Nat. Bur. Stand. (U.S.), Tech. Note 355 (1967).

18. Hill, J. E., and Didion, D. A., Comparative performance of two postal service vehicles operated on gasoline, compressed natural gas, and propane, ASME paper 74-WA/HT-26, presented at the ASME winter annual meeting, New York, N.Y. (Nov. 17-22, 1974); see also, Hill, J. E., and Didion, D. A., Comparison of power output and exhaust pollutants of two postal service vehicles operated on three hydrocarbon fuels, Nat. Bur. Stand. (U.S.), NBS-IR 74-460 (Sept. 1973).

19. Sparks, L. L., R. L. Powell, and W. J. Hall, Reference tables for low-temperature thermocouples, Nat. Bur. Std's. (U.S.), NBS Monograph 124 (June 1972).

Appendix A.  Physical Properties of $H_2$ and $CH_4$

| Physical Property | $p-H_2$ | $CH_4$ |
|---|---|---|
| Molecular weight | 2.016 | 16.043 |
| Triple point pressure, atm | 0.0695 | 0.1159 |
| Triple point temperature, K | 13.803 | 90.680 |
| Normal boiling point (NBP) temperature, K | 20.268 | 111.632 |
| Critical pressure, atm | 12.759 | 45.355 |
| Critical temperature, K | 32.976 | 190.53 |
| Critical density, $g/cm^3$ | 0.0314 | 0.1628 |
| Liquid density @ NBP, $g/cm^3$ | 0.0708 | 0.4226 |
| Vapor density @ NBP., $g/cm^3$ | 0.00134 | 0.00182 |
| Solid density @ triple point, $g/cm^3$ | 0.0865 | 0.4872 |
| Gas density @ NTP, $g/m^3$ | 83.764 | 651.19 |
| Density ratio: NBP liquid-to-NTP gas | 845 | 649 |
| Heat of fusion, J/g | 58.23 | 58.47 |
| Heat of vaporization @ NBP, J/g | 445.59 | 509.88 |
| Heat of combustion (low), kJ/g | 119.96 | 50.02 |
| Limits of flammability in air, vol. % | 4 to 75 | 5 to 15 |
| Limits of detonability in air, vol. % | 18 to 59 | -- |
| Stoichiometric composition in air, vol. % | 29.53 | 9.48 |
| Minimum energy for ignition in air, J | 0.02 | 0.29 |
| Ignition temperature, K | 858 | 810 |
| Flame temperature in air, K | 2318 | 2148 |
| Percentage of thermal energy radiated to surroundings from burning liquid pool, % | 25 | 23 |
| Flame velocity in NTP air, cm/s | 265 | 39 |
| Quenching gap in NTP air, cm | 0.06 | 0.22 |

---

NBP = normal boiling point

NTP = 1 atm and 293.15 K

## Appendix B.  Use of Figure 23

For this example a passenger vehicle similar to the one tested in
this report will be chosen; a vent area of 50 $cm^2$ is assumed.  As one
progresses from zero to increasing leakage flow rates (on figure 23)
the first curve encountered is labeled "Preferred Vent Areas (low risk)".
The corresponding leakage flow rate (with $H_2$ or $CH_4$) is approximately
9 sccs at 50 $cm^2$ vent area.  This means that a sustained leakage flow
rate of 9 sccs will produce combustible gas concentrations in the
passenger compartment which are $\leq$ 20% of LEL.

These statements are true only for steady-state leakage and stated
concentration levels are attained after equilibrium conditions have
been established.

As the leakage flow rate is increased, maintaining 50 $cm^2$ of vent
area, the next set of curves encountered on figure 23 are labeled
"Minimum Vent Area (high risk)".  The $H_2$ and $CH_4$ curves are almost
coincident at this point.  Leakage flow rate is approximately 69 sccs
for either $H_2$ or $CH_4$; this means that under equilibrium conditions a
concentration level of 100% of LEL can be expected at sensor 5 in the
passenger compartment.

When the leakage flow rate exceeds 69 sccs for the same vent area
(50 $cm^2$), the volume of combustible gas ($\geq$ 100% of LEL) inside of the
vehicle increases.  The curve labeled "Explosion hazard for entire upper
region of car interior" constitutes an arbitrarily chosen boundary at
which the region of the car interior from head-level to ceiling is
occupied by combustible gas ($\geq$ 100% of LEL).  This limit is reached
at approximately 78 sccs for $CH_4$ and 117 sccs for $H_2$.

If increased leakage flow rates are realized (maintaining 50 $cm^2$
vent area) we go from the shaded to the unshaded region on figure 23.
Therefore, an increasing portion of the vehicle interior volume is
being filled with combustible gas until, possibly, the total volume is
flammable ($\geq$ 100% of LEL).

. TITLE AND SUBTITLE

Efflux of Gaseous Hydrogen or Methane Fuels

from the Interior of an Automobile

. PERFORMING ORGANIZATION NAME AND ADDRESS

NATIONAL BUREAU OF STANDARDS
DEPARTMENT OF COMMERCE
WASHINGTON, D.C. 20234

. Sponsoring Organization Name and Complete Address *(Street, City, State, ZIP)*

Mr. R. L. Ullrich
Director, Motor Equipment Research & Technology Division
FZR; GSA-FS (FZR)
Crystal Mall Bldg. 4, Room 323, Washington, D.C. 20406

reau of Standards, Department of

. ABSTRACT *(A 200-word or less factual summary of most significant information. If document includes a significant bibliography or literature survey, mention it here.)*

Gasoline-powered automobiles are being converted to operate on gaseous fuels such as $H_2$ or $CH_4$. These fuels are commonly stored in containers located in the trunk of the car. Potential leakage of these gaseous fuels into the passenger compartment of the vehicle constitutes a safety threat. Definitive experiments were performed to identify the explosion hazards, establish venting criteria and obviate general safeguards for $H_2$ or $CH_4$ fueled passenger vehicles. Appropriately designed ventilation systems significantly reduce the safety hazards associated with accumulated combustible gases. Vents are recommended for all autos <u>converted</u> to burn $H_2$ or $CH_4$ and may possibly be eliminated in new cars that are <u>designed</u> for gaseous fuel operation. Combustible gas warning systems are recommended, at least in the interim, for all (converted and new-design) gaseous fueled vehicles. $H_2$ and $CH_4$ gases appear equally safe as vehicular fuels if used in properly designed vehicles.

7. KEY WORDS *(six to twelve entries; alphabetical order; capitalize only the first letter of the first key word unless a prop name; separated by semicolons)*

\utomobile; detection; dispersion; explosion; fire; hydrogen; leakage; methane; ;afety; vents.

8. AVAILABILITY          ☒ Unlimited

☐ For Official Distribution. Do Not Release to NTIS

☐ Order From Sup. of Doc., U.S. Government Printing Office
Washington, D.C. 20402, SD Cat. No. C13

☐ Order From National Technical Information Service (NTIS)
Springfield, Virginia 22151

# NBS TECHNICAL PUBLICATIONS

## PERIODICALS

**JOURNAL OF RESEARCH** reports National Bureau of Standards research and development in physics, mathematics, and chemistry. It is published in two sections, available separately:

● **Physics and Chemistry (Section A)**
Papers of interest primarily to scientists working in these fields. This section covers a broad range of physical and chemical research, with major emphasis on standards of physical measurement, fundamental constants, and properties of matter. Issued six times a year. Annual subscription: Domestic, $17.00; Foreign, $21.25.

● **Mathematical Sciences (Section B)**
Studies and compilations designed mainly for the mathematician and theoretical physicist. Topics in mathematical statistics, theory of experiment design, numerical analysis, theoretical physics and chemistry, logical design and programming of computers and computer systems. Short numerical tables. Issued quarterly. Annual subscription· Domestic, $9.00; Foreign, $11.25.

**DIMENSIONS/NBS** (formerly Technical News Bulletin)—This monthly magazine is published to inform scientists, engineers, businessmen, industry, teachers, students, and consumers of the latest advances in science and technology, with primary emphasis on the work at NBS. The magazine highlights and reviews such issues as energy research, fire protection, building technology, metric conversion, pollution abatement, health and safety, and consumer product performance. In addition, it reports the results of Bureau programs in measurement standards and techniques, properties of matter and materials, engineering standards and services, instrumentation, and automatic data processing.

Annual subscription: Domestic, $9.45; Foreign, $11.85.

## NONPERIODICALS

**Monographs**—Major contributions to the technical literature on various subjects related to the Bureau's scientific and technical activities.

**Handbooks**—Recommended codes of engineering and industrial practice (including safety codes) developed in cooperation with interested industries, professional organizations, and regulatory bodies.

**Special Publications**—Include proceedings of conferences sponsored by NBS, NBS annual reports, and other special publications appropriate to this grouping such as wall charts, pocket cards, and bibliographies.

**Applied Mathematics Series**—Mathematical tables, manuals, and studies of special interest to physicists, engineers, chemists, biologists, mathematicians, computer programmers, and others engaged in scientific and technical work.

**National Standard Reference Data Series**—Provides quantitative data on the physical and chemical properties of materials, compiled from the world's literature and critically evaluated. Developed under a world-wide

program coordinated by NBS. Program under authority of National Standard Data Act (Public Law 90-396).

NOTE. At present the principal publication outlet for these data is the Journal of Physical and Chemical Reference Data (JPCRD) published quarterly for NBS by the American Chemical Society (ACS) and the American Institute of Physics (AIP). Subscriptions, reprints, and supplements available from ACS, 1155 Sixteenth St. N. W., Wash. D. C. 20056.

**Building Science Series**—Disseminates technical information developed at the Bureau on building materials, components, systems, and whole structures. The series presents research results, test methods, and performance criteria related to the structural and environmental functions and the durability and safety characteristics of building elements and systems.

**Technical Notes**—Studies or reports which are complete in themselves but restrictive in their treatment of a subject. Analogous to monographs but not so comprehensive in scope or definitive in treatment of the subject area. Often serve as a vehicle for final reports of work performed at NBS under the sponsorship of other government agencies.

**Voluntary Product Standards**—Developed under procedures published by the Department of Commerce in Part 10, Title 15, of the Code of Federal Regulations. The purpose of the standards is to establish nationally recognized requirements for products, and to provide all concerned interests with a basis for common understanding of the characteristics of the products. NBS administers this program as a supplement to the activities of the private sector standardizing organizations.

**Federal Information Processing Standards Publications** (FIPS PUBS)—Publications in this series collectively constitute the Federal Information Processing Standards Register. Register serves as the official source of information in the Federal Government regarding standards issued by NBS pursuant to the Federal Property and Administrative Services Act of 1949 as amended, Public Law 89-306 (79 Stat. 1127), and as implemented by Executive Order 11717 (38 FR 12315, dated May 11, 1973) and Part 6 of Title 15 CFR (Code of Federal Regulations).

**Consumer Information Series**—Practical information, based on NBS research and experience, covering areas of interest to the consumer. Easily understandable language and illustrations provide useful background knowledge for shopping in today's technological marketplace.

**NBS Interagency Reports (NBSIR)**—A special series of interim or final reports on work performed by NBS for outside sponsors (both government and non-government). In general, initial distribution is handled by the sponsor; public distribution is by the National Technical Information Service (Springfield, Va. 22161) in paper copy or microfiche form.

Order NBS publications (except NBSIR's and Bibliographic Subscription Services) from: Superintendent of Documents, Government Printing Office, Washington, D.C. 20402.

## BIBLIOGRAPHIC SUBSCRIPTION SERVICES

The following current-awareness and literature-survey bibliographies are issued periodically by the Bureau:
Cryogenic Data Center Current Awareness Service

A literature survey issued weekly. Annual subscription: Domestic, $20 00; foreign, $25.00.

Liquefied Natural Gas. A literature survey issued quarterly. Annual subscription $20 00.

Superconducting Devices and Materials. A literature

survey issued quarterly. Annual subscription: $20.00. Send subscription orders and remittances for the preceding bibliographic services to National Technical Information Service, Springfield, Va. 22161.

Electromagnetic Metrology Current Awareness Service Issued monthly Annual subscription. $100 00 (Special rates for multi-subscriptions). Send subscription order and remittance to Electromagnetics Division, National Bureau of Standards, Boulder, Colo. 80302.